Python for Mechanical and Aerospace E...

by Alex Kenan

Learn some of Python's most popular libraries like BeautifulSoup, Matplotlib, Numpy, Requests, and Tkinter by scraping a website for aluminum alloy information, graphing satellite orbits, creating a GUI to convert different units, plotting airfoil coordinates, publishing programs to PDF, and more!

https://www.alexkenan.com/pymae/

Python for Mechanical and Aerospace Engineering

Ebook ISBN 978-1-7360606-0-5

Paperback ISBN 978-1-7360606-1-2

Hardcover ISBN 978-1-7360606-3-6

Version 1.06

Copyright © 2020 Alexander Kenan. Some Rights Reserved. This work is licensed under a Creative Commons Attribution – NonCommercial – ShareAlike 4.0 International License. See http://creativecommons.org/licenses/by-nc-sa/4.0/ and **The End** chapter for more information.

https://www.alexkenan.com/pymae/

pythonforengineering@gmail.com

Table of Contents

CHAPTER 1: INTRODUCTION TO PYTHON — 5

CHAPTER 1.5: FIZZBUZZ — 39

CHAPTER 2: GRAPHING THRUST AVAILABLE AND THRUST REQUIRED — 45

CHAPTER 3: GRAPHING DYNAMIC PRESSURE DURING A ROCKET LAUNCH — 64

CHAPTER 4: GETTING AND PLOTTING AIRFOIL COORDINATES — 82

CHAPTER 5: MODELING A 2-BODY ORBIT IN 2D AND 3D — 99

CHAPTER 6: UNIT CONVERSIONS — 124

CHAPTER 7: INTRODUCTION TO WEB SCRAPING — 148

CHAPTER 8: MODELING CAMERA SHUTTER EFFECT — 165

CHAPTER 9: WRITING REPORTS WITH PWEAVE — 175

THE END — 188

ABOUT THE AUTHOR — 189

LEGAL STUFF — 191

Chapter 1: Introduction to Python

What do you think of when you hear the word "programming?" For most of the general population, programming conjures up images of advanced mathematics, smelly geeks, and "hacking into the mainframe," whatever that means. For engineers and engineering students, programming generally brings back memories of triumphs and struggles of introduction to computer science courses.

These traditional computer science courses for engineering focus on the fundamentals of programming without demonstrating the wide array of practical applications for outside fields. As a result, engineering students know (or, at least, learned) the principles of inheritance and object-oriented programming, but the courses fail to link these concepts to actual engineering applications. As a result, mechanical and aerospace engineering disciplines get pushed to MATLAB because the computational abilities of programming languages tend to not be highlighted. MATLAB has traditionally dominated the engineering space because it is viewed as a batteries-included software kit that is focused on functional programming. Everything in MATLAB is some sort of array, and it lends itself to engineering integration with its toolkits like Simulink and other add-ins. The downside of MATLAB is that it is proprietary software, the license is expensive to purchase, and it is more limited than Python for doing tasks besides calculating or data capturing.

Python is a general-purpose programming language that can be used to write code for both small and large projects. Companies like Dropbox, Instagram, Instacart, and reddit, among others, use Python in some or all of their production code. The language has three major selling points: it is free, it has good code readability, and it has a batteries-included philosophy, meaning that it can be quickly downloaded and used. The majority of the time, Python code reads very similarly or almost exactly like normal English. That helps turn your code from pseudocode, where you get the general idea of what you want your program to do, to executable code. Python is also highly regarded because of its wide-ranging standard library. A library, also called a module, is a folder of pre-written code that you can use and re-use. Using a Python module is like calling your favorite car guy/gal to help diagnose a problem with your car. You get to benefit from his/her expertise without having to do the hard work of researching the problem, ordering parts, diagnosing the problem, etc. For a Python example, you can use the built-in operating system module *os* to help you move a file from your Documents folder to your Desktop folder. All you have to do is tell Python which file to move, and to where, and it takes care of the rest. The standard library has modules to help manage files on your computer, calculate and manage dates and times, time how long it takes your code to run, help test your code, and many more.

    Another benefit of Python is that it is open-source, which makes it easy to use others' work to help you. While Python has a good standard library, it has even better third-party library support and distribution system; there is a standard package management system called "pip" that downloads libraries with one command: *pip install library_name*. You can take advantage of extensive numerical methods with the library Numpy (Numerical Python, pronounced numb-pie), and you don't have to worry about the inner workings of the programs you are using. Numpy has a guide [1] that converts MATLAB syntax to Python syntax for any MATLAB converts that may be reading.

A common saying is that "if you feed someone fish, they will eat for a day. If you teach them to fish, they will eat for a lifetime." Learning how to program in Python is not teaching someone how to fish; it is teaching them how to decide which way of acquiring food is the best. Python might not always be the best tool to accomplish something, but it is almost always second-best. And in this case, being second-best at a lot of things is definitely better than being the best in a very narrow field. Being a jack-of-all trades means that Python can be used to record or fetch data, process it, display it, save it, publish it to a website, and email it, which makes it much more versatile than MATLAB.

This book assumes a college junior level of mechanical/aerospace engineering understanding with examples that touch on thrust available and thrust required for an aircraft, dynamic pressure and how it changes with altitude and velocity, airfoils, orbital mechanics and orbital parameters, and mechanical properties of different aluminum alloys. Don't be scared if you don't understand all of these topics; a complete understanding of these subjects is not required. The Python examples will hit home more clearly if you understand the engineering subject that is being used for the example. Showing these engineering examples takes the abstract philosophy of object-oriented programming and turns it into an actual project. For me, I finally understood an orbital parameter only when I was able to see it in 2D and 3D, which we will learn in Chapter 5.

This book does not assume any level of programming experience. This chapter will show you the basic structure of Python, its default datatypes, and the rules of programming in Python. It is by no means a comprehensive tutorial on computer science principles or philosophy. There can be semantic differences between how fast a particular datatype is compared to another datatype, but we won't be making a big deal out of any of that. We are not here to make something run 0.00003 seconds faster. We are here to make cool engineering applications in Python. We will also generally try to follow Python best practices, but there may be one or two coding practices that are "discouraged" but improve clarity/understanding or are easier for us to write.

One last thing: this book is not going to be a good beach book. This book is intended to be used as a practical guide; in order to get the most out of it, you need to program the examples and follow along. Learning to program is a lot like learning math: as much as you try, there is no substitute for practice. In order to practice, you need to download Python.

Downloading Python

There are several different ways to download Python. I highly recommend downloading Python via the Anaconda distribution because Anaconda bundles the popular third party library "science" stack of Matplotlib, Numpy, and Scipy, among others.

The application that we will use within the Anaconda distribution is called Spyder. Spyder is called an Integrated Development Environment, or IDE. IDEs have lots of tools on top of the normal text editor area. An IDE is similar to a word processor like Microsoft Word or Apple Pages. You *could* write a book or paper in TextEdit or Notepad with just a bare-bones text editor, but the word processors have powerful features like spelling and grammar editing, advanced formatting options, etc. In particular, Spyder has two additional features that make it useful: a display that shows the name and value of all active variables (called the Variable Explorer), and a console/notebook window. The Variable Explorer is a handy troubleshooting tool that shows you what variables you have created and what their values are, and the console is the Python interpreter that takes your Python code and runs it. You can use the console to test out snippets of code before putting them into your main program. Spyder also has a pre-formatted layout that mimics the standard MATLAB layout.

Let's walk through downloading Spyder. Go to the download link at https://www.anaconda.com/distribution/ and download the Python 3.x version of Anaconda Navigator. At the time of writing, this version is Python 3.8. This book will use Python 3 for all examples. Python 3 was released in 2008 and superseded Python 2. The new Python 3 was not entirely backwards-compatible with Python 2; some code written in Python 2 had to be edited to work in Python 3. Because of this backwards compatibility issue, some users and companies have been stalwart holdouts clinging to Python 2. This has sparked strong debate in the Python community that is not really of concern to us. If you are learning Python, learn Python 3.

Open Anaconda and let it run any updates. This is what Anaconda Navigator should look like when it is ready.

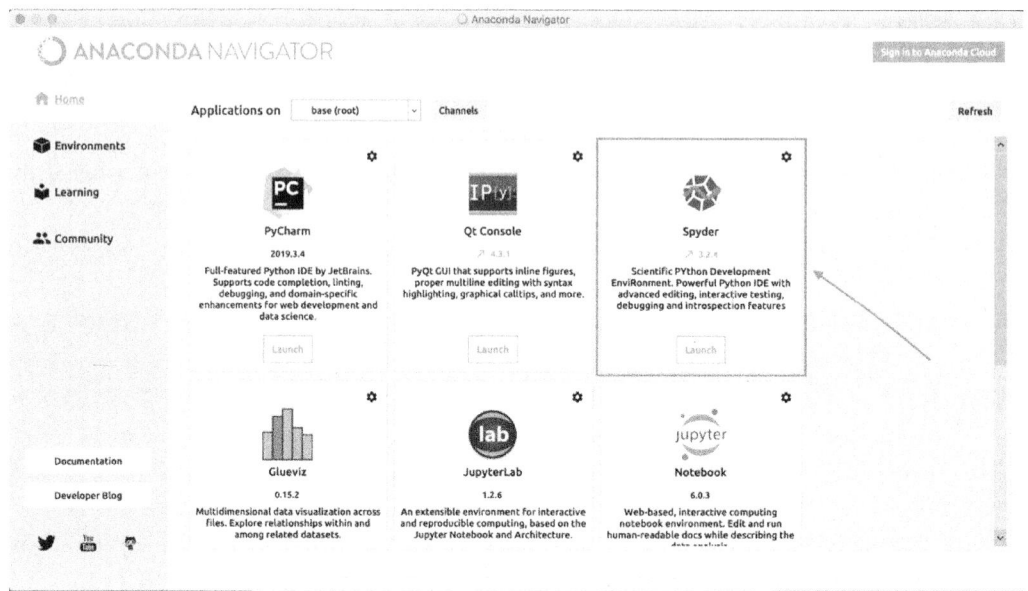

Click on "Launch" under Spyder. When I first downloaded Anaconda, I was getting lots of errors trying to launch Spyder. I ultimately had to switch from "Application on base (root)" to "Applications on anaconda3" to open Spyder.

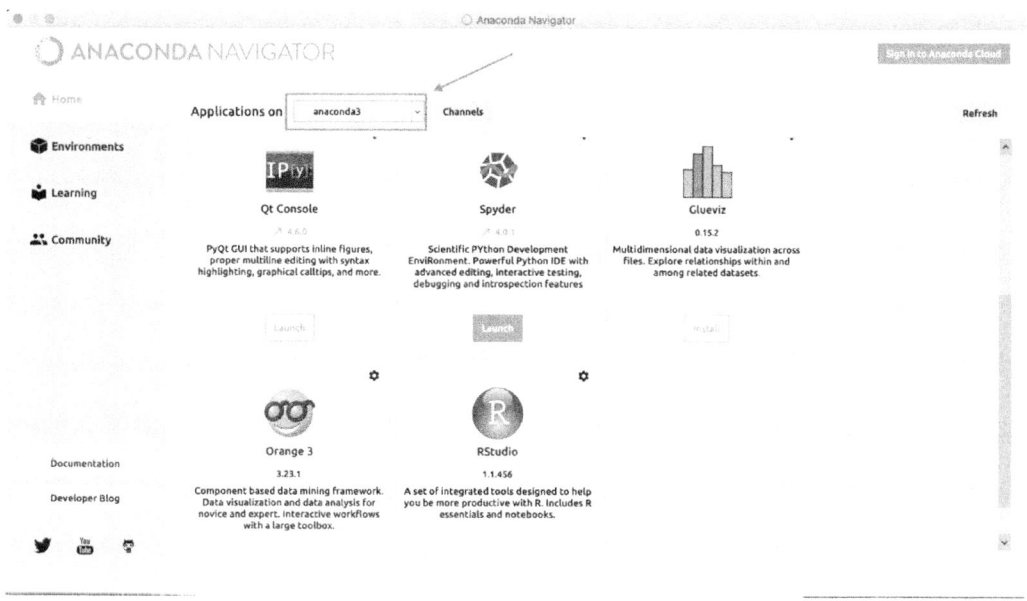

The three areas of Spyder are highlighted below. We will use the terms "console" and "notebook" interchangeably. Both terms refer to the place where you can input Python code and run it immediately. Make sure to click the "Variable explorer" tab in the Variable Explorer area so that you can actually see the variables.

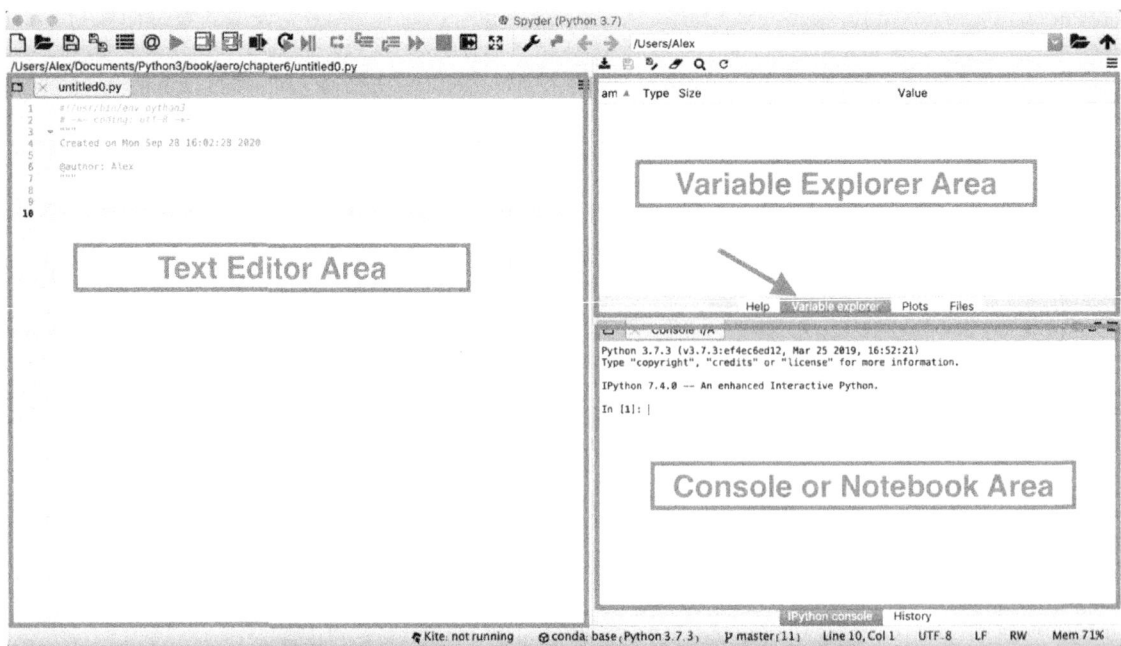

I ended up changing the layout settings to be light themed, but use whatever style you prefer. We will primarily be writing code in the text editor area and "printing" results to the console or notebook area.

Downloading Python Packages

Pip is the package management system that allows users to publish, distribute, and download third-party libraries. It can download any external libraries that have been published to PyPI, the Python Packaging Index. You interact with pip via the command line of a terminal, which is the CMD application for Windows or the Terminal application for Mac/Linux. Use the following to verify that pip is working on your computer:

- On **Windows**, open Anaconda Prompt, which is a stand-alone application that is created when you download Anaconda. This should open a CMD terminal that already points to the Anaconda folder. Type *pip install numpy*, hit Enter, and you should see something like *Requirement already satisfied: numpy in ./opt/anaconda3/lib/python3.8/site-packages (1.18.1)*.

- On **Mac/Linux**, open Terminal, type *which pip*, and hit Enter. You should see something with *anaconda3* in the path.
    - If you do see that, type *pip install numpy* and that command should return *Requirement already satisfied: numpy in ./opt/anaconda3/lib/python3.8/site-packages (1.18.1)*.
    - If you don't see something with *anaconda3* in the path, you will have to manually point to pip. On my computer, Anaconda3 is installed in */Users/Alex/anaconda3/*. Pip is located within the */bin/* folder of */anaconda3/*. In order to call pip, you need to specify the full path like */Users/Alex/anaconda3/bin/pip install numpy*. That command should get you to *Requirement already satisfied: numpy in ./opt/anaconda3/lib/python3.8/site-packages (1.18.1)*.

If you see anything requiring pip in the **Import** section of any chapter, you will need to return to the above instructions in order to install the required libraries.

Python Basics

This book will go through some Python basics before moving into code examples. I have always learned the most about Python by trying to write programs instead of just reading about it. Follow along, tweak different parts of the programs in the following chapters, and don't be afraid to think "Huh, that's weird..." or "I wonder what would happen if..." or "I wonder if I can do [x] in Python..." (Hint: yes, you probably can.) If you feel that this chapter moves too quickly, I would recommend trying a free tutorial resource like Codecademy or Udemy to get some of the basics down. Thanks to a Freedom of Information Act request, the National Security Agency (NSA) declassified and published its 595 page Introduction to Python course syllabus. This document, and the rest of this book's example programs, can be found at this book's GitHub repository [2]. In this section, we will explain some Python basics, and then in the next chapter, we can begin coding. This chapter will be the longest one so that the Python basics can be adequately covered. Feel free to skip to the next section if you are ready to see some programming in action, but I would recommend at least briefly reviewing the rest of this section to understand the syntax and language.

Everything in Python is some sort of datatype. It's similar to how everything in the world is some type of thing (or has a classification), like a plant, an animal, a car, or a star. It's hard to explain and conceptualize, but easy to understand. Numbers are the easiest to explain: the first datatype is *int*, short for integer. An integer is any whole number such as 1, 7, -15, 15415, 801597, and all of the other whole, non-decimal numbers. In Python, you can perform the four basic mathematical operations of addition, subtraction, multiplication, and division with their usual operators. Addition is represented with +, subtraction is represented with -, multiplication is represented with *, and division is represented with /. Go to Spyder, click in the Notebook area by the line that says *In[1]:*, type *1+3*, and hit Enter on your keyboard. You should see the following:

```
In [1]: 1+3
Out[1]: 4

In [2]:
```

Congratulations! You just ran your first piece of Python code. Instead of using screenshots all of the time, text representations will be shown for most code. For example, the above screenshot is equivalent to:

```
In[1]: 1+3
Out[1]: 4
```

If we were using a real console instead of the Notebook, the screenshot would look like this:

```
>>> 1+3
4
```

A "real console" is the Python interpreter without the Spyder IDE. On Windows, if you open Anaconda Prompt, type *python3*, and hit Enter, you will have a console open.

15

Likewise, on Mac, open Terminal, type *python3*, and hit Enter. The two visual representations are identical. It is easier for beginners to use the Notebook because it is built into Spyder and integrates with the Variable Explorer.

Notice that there is no semi-colon or anything else at the end of a line of Python code. Go ahead and try out subtraction, multiplication, and division. Using division leads to an interesting discovery:

```
>>> 10/5
2.0
```

We have just met another datatype: the *float*. You can think of *float* as a floating decimal point number, or a decimal. You will need to be careful if you ever mix *int* and *float* because their behavior can be different. Generally, if you mix the two datatypes, the result is going to be a *float*. You can perform all of the same math functions on *floats*. Try performing the basic math functions with a *float* and an *int* together. You can also put spaces between the numbers and the operation symbol, and Python won't get mad:

```
>>> 4*2.0
8.0
>>> 4 * 2.0
8.0
```

If you really want to get an *int* as a result, you can cast, or convert, from one type to the other by doing this:

```
>>> int(8.0)
8
```

By converting a *float* to an *int*, we are essentially calling a function, which we will talk more about in the next chapter. Be careful: casting as an *int* will round towards zero.

```
>>> int(8.9999)
8
>>> int(-8.9999)
-8
```

Let's go through some of the remaining math operators that we haven't talked about: exponentiation, remainder, floor division (which computes the division and returns the rounded-down number), and rounding.

```
>>> 2**4   # exponentiation
16
>>> 15 % 6   # remainder or "modulus"
3
>>> 20 // 7   # floor division
2
>>> round(20/7, 3)   # rounding
2.857
```

So, you can use the *round()* function to do any rounding before you cast a *float* as an *int*.

17

Did you see the # after the code? The # is a single-line comment; it means that anything after that symbol on the same line will not be executed. You can use comments to write notes to yourself, temporarily "comment out" parts of code to troubleshoot or test, write reminders for later, or make your code easier for others to follow. Triple apostrophes (''') or triple quotes (""") can be used to create multi-line comments.

```
# single line comment
# single line comment that happens to
#            spill over onto the next line
"""
this is a
m
 u
  l
   t
    i
line comment
"""
```

There is also a *complex* type for complex numbers that uses *j* for the imaginary number.

```
>>> x = 1 + 3j   # create a complex number
>>> type(x)   # request x's datatype
<class 'complex'> # or complex
```

Let's move on to text. Text is called *str*, short for string. We represent a *str* by wrapping it around either a single quote ' or a double quote ". It does not matter which convention you use as long as you are consistent for that specific time.

```
>>> 'this is a string'
```

18

```
'this is a string'
>>> "this is also a string"
"this is also a string"
>>> "this will return an error'
  File "<ipython-input-10-14355ec08dd5>", line 1
    "this will return an error'
                               ^
SyntaxError: EOL while scanning string literal
```

Congratulations, you have encountered your first error or bug! If Python doesn't understand what you are asking it to do, it will raise an *Error* (sometimes called an *Exception*). Remember, the computer is going to do what you told it to do, not necessarily what you wanted it to do. In order to troubleshoot these errors, you need to carefully read the traceback – what the error says – to figure out what happened. In this case, it says that it ran into a `SyntaxError` on line 1: `EOL (end of line) while scanning string literal`. The interpreter also tried to use ^ to show us approximately where the error was in the line. So, Python reached the end of the line and did not think the line was complete. We mentioned above how mixing the two `str` conventions will get you into trouble. If we are consistent, the error will go away:

```
>>> "this will no longer return an error"
'this will no longer return an error'
```

If you need to mix ' and ", you can use the escape character \ to "escape" the character you don't want to be interpreted as the beginning or end of the `str`.

```
>>> "you could to this to avoid ' causing problems"
>>> 'or you could do this and quote "as many times as you want" '
```

```
>>> 'or you could use contractions as much as you\'d like'
```

Now is a good time to learn about assigning variables. You can store anything as a variable to access it later. Variable assignment is performed using the = operator.

```
>>> x = 4 + 5
>>> x
9
>>> x + 5
14
```

Variable names must not start with a number, must not contain spaces, and cannot be the same name as an existing keyword (names like *with, open, in, True, False, and, or, import, from, continue, yield, try, except, finally, def, pass, is, round, max, abs, print, id, class, bool, int, float, list, str, dict, set, if, elif, else* and others). Basically, if Spyder puts bold formatting around a variable name, you are using a Python keyword, and you should pick a different variable name. Python convention is to use all lowercase letters with underscores instead of spaces *like_this*. The use of *CamelCase* or *camelCase* is discouraged but not forbidden. Here are some valid and invalid variable names:

Valid	Invalid

x list_of_values range_of_something	id list range of something

Python allows chained variable assignment, so you can do something like:

```
>>> force_on_obj1 = force_on_obj2 = force_on_obj3 = 34
# or
>>> force_on_obj1, force_on_obj2 = 34, 35
```

It is also important to note that Python executes the line of code with the information that it knows at the time. If you were to set two variables equal to each other, then change one variable, the other will still retain the original value.

```
>>> speed_of_car = 5
>>> speed_of_second_car = speed_of_car
>>> speed_of_car = 7
>>> speed_of_second_car
5
```

This behavior is a little different for *lists*, which we will explore later in this chapter.

You can also use shorthand for the simple math and *str* assignments:

```
>>> x = x + 1
>>> x += 1    # these statements are equivalent
```

It works for + - * and / operators. On *strs*, you would mostly use the += operator to add to a *str*:

21

```
>>> string = 'hi'
>>> string += '; well hello to you too'
>>> string
'hi; well hello to you too'
```

Back to our *str* type, the *len()* function returns the length of any *str* as an *int*. It will count everything in the *str*: whitespace, letters, numbers, punctuation, etc.

```
>>> len("I don't know how many characters this is")
40
```

When we pass a *str* to the *len()* function, we are passing a "parameter" or an "argument." An argument is a value that is passed to the function. In the example above, the argument is the *str*, and the *len()* function returns the length of the *str*. Functions can have no arguments, many arguments, and anything in between.

Another powerful *str* characteristic is being able to slice the *str* and get the character or characters at any location. It uses the syntax *string_variable[index]*, where *string_variable* is the variable name, and *index* is the integer number, counting from the left, that equates to the position within that *str*. We can use *str* slicing to try to access the first letter in *Quick brown fox* shown below:

```
>>> text = 'Quick brown fox'
>>> text[1]
'u'
```

We have encountered another Python ideological battleground: zero-indexing. Python is zero-indexed, meaning the first letter of *text*, *Q*, corresponds to index=0.

Index=1 is actually the second letter, *u*. Most programming languages are zero-indexed, meaning that the "first" letter in a *str* corresponds to the "zeroth" index. More calculation-oriented programs are one-indexed, meaning that the position corresponds with the index (e.g. the "first" letter in a *str* corresponds to the first index.) MATLAB, for example, is one-indexed. It is best to just accept this reality rather than argue about which is correct. The indices of the characters in *text* are as follows:

0	1	2	3	4	5	6	7	8	9	10	11	12	13	14
Q	u	i	c	k		b	r	o	w	n		f	o	x

```
>>> len(text)
15
```

This means that trying to get the last character in *text* with *text[15]* will return an error.

```
>>> text[15]
Traceback (most recent call last):
    File "<stdin>", line 1, in <module>
IndexError: string index out of range
```

Str slicing also works with negative numbers. The index *-1* corresponds to the last character, *-2* corresponds to the second-to-last character, etc. *text[-1]* is generally used to get the last character in a *str*, not *text[len(text)-1]* even though both return the same result.

23

-15	-14	-13	-12	-11	-10	-9	-8	-7	-6	-5	-4	-3	-2	-1
Q	u	i	c	k		b	r	o	w	n		f	o	x

Notice that we can pass a yet-to-be-calculated operation as a parameter. We can pass *text[len(text)-1]* and still get the same result as if we had passed *text[14]*. There is a particular subset of programmers that pride themselves in writing the most code possible on a single line, much to the detriment of readability.

Another useful method is the *upper()* function, which capitalizes any letters in the *str*.

```
>>> text = 'Quick brown fox'
>>> text.upper()
'QUICK BROWN FOX'
>>> text
'Quick brown fox'
```

Notice that, by itself, *upper()* does not change the *str* in-place. If you want to make *text* all uppercase, you need to re-assign the *text* variable.

```
>>> text = text.upper()
>>> text
'QUICK BROWN FOX'
```

It is generally good practice to create a new variable instead of modifying an existing variable in-place. The "best practice" version of the above code looks like this:

```
>>> text_upper = text.upper()
```

This practice helps prevent confusion and unexpected downline effects later in code.

There are many other useful *str* functions; some of the most useful are included below.

Str function	What it does	Example
| len(*str*) | Returns the length of the *str* | ```
>>> x = 'hello'
>>> len(x)
5
``` |
| upper() and lower() | Makes all letters uppercase or lowercase, respectively | ```
>>> x.upper()
'HELLO'
``` |
| *str* slicing | Returns part of the *str* by the index | ```
>>> x[0]
'h'
>>> x[-1]
'o'
>>> x[2:]
'llo'
>>> x[:2]
'he'
``` |
| format(*str*) | Formats the *str* (usually used with the *print()* function) | ```
>>> print('{} there'.format(x))
'hello there'
``` |
| replace(*str1*, *str2*) | Replaces *str1* with *str2* | ```
>>> x.replace("l", "p")
"heppo"
``` |
| count(*str1*) | Counts the number of occurrences *str1* | ```
>>> x.count('e')
1
``` |
| strip() | Removes any leading or trailing whitespace. You can use *lstrip()* for removing left whitespace (leading) or *rstrip()* for any right whitespace (trailing) | ```
>>> y = '   hello   '
>>> y.strip()
'hello'
``` |
| join(*str1*) | Creates a new *str* that joins the contents of the old *str*, one character at a time, by using the supplied *str1* as a separator | ```
>>> ','.join(x)
'h,e,l,l,o'
>>> ' '.join(x)
'h e l l o'
``` |

| index(*str1*) | Returns the index of the argument *str1* | ```
>>> x.index('e')
1
>>> x.index('elL')
1
``` |
```

Str methods are an instance where it can be more readable to stack multiple operations together:

```
>>> text = 'Quick brown fox'
>>> text = text.upper().replace(' ', '_').lower()
>>> text
'quick_brown_fox'
# is the same as
>>> text = 'Quick brown fox'
>>> text = text.upper()
>>> text = text.replace(' ', '_')
>>> text = text.lower()
>>> text
'quick_brown_fox'
```

The next datatype is the *List*. A *List* can be a collection of anything; think of it as a one-dimensional array. It has one "row" but as many "columns" as you want. Usually the datatypes are the same in a given *List*, but that convention does not have to be followed. *Lists* can be sliced just like *strs*. The official datatype name is *List* or *[ ]*, so don't name any list variable plain old *List*!

```
>>> list_of_nums = [1, 2, 3, 4, 5]
>>> list_of_nums[0]
1
>>> len(list_of_nums)
5
```

You can add to *Lists* by using the *append()* method, which accepts one argument: the variable you want to add to the *List*.

```
>>> list_of_nums.append(7)
>>> list_of_nums
[1, 2, 3, 4, 5, 7]
```

A *List* is almost exactly like an array type in MATLAB. There are some small differences between the two datatypes, but you can consider *Lists* and *arrays* to be the same.

Below is another list of some of the most useful *List* functions:

List function	What it does	Example

Len(y)	Returns the length of *List* y	```>>> x = ['a', 'b', 'c']```
		```>>> Len(x)```
		```3```

append(y)	Adds y to the end of the existing *List*	```>>> x.append('d')```
		```>>> x```
		```['a', 'b', 'c', 'd']```

List slicing	Returns part of the *List* by the index	```>>> x[0]```
		```'a'```
		```>>> x[-1]```
		```'d'```
		```>>> x[2:]```
		```['c', 'd']```
		```>>> x[:2]```
		```['a', 'b']```

insert(i, y)	Inserts y at index i in the *List*	```>>> x.insert(2, 'f')```
		```>>> x```
		```['a', 'b', 'f', 'c', 'd']```

remove(y)	Removes the first value in the *List* that equals y. Returns a *ValueError* if nothing equals y.	```>>> x.remove('f')```
		```>>> x```
		```['a', 'b', 'c', 'd']```

count(y)	Counts the number of occurrences of y in a *List*	```>>> x.count('a')``` ```1```

pop(i)	Removes the value at index i. Returns that value	```>>> y = x.pop(0)```
		```>>> y```
		```'a'```
		```>>> x.insert(0, 'a')```
		```>>> x.pop(0)```
		```x```
		```['b', 'c', 'd']```

| index(y, optional start, optional end) | Returns the first index of y in the list. Can use optional arguments start and end for starting and ending indices | ```
>>> x.index('d')
3
``` |
| enumerate(y) | Gives a *tuple* of (index, item) for a *list*. | ```
>>> y = [11, 22, 33]
>>> for index, item in enumerate(y):
>>>     print(index, item)
0 11
1 22
2 33
``` |
| copy() | Copies the *list*. This is the best way to copy a *list*. | ```
>>> list1 = [1, 2, 3, 4]
>>> list2 = list1.copy()
``` |

The next datatype is a *tuple*. A *tuple* is essentially an immutable (unchangeable) *list*. A *tuple* is created with parentheses *()*. Once a *tuple* exists, nothing about it can be changed: neither its length, nor any of its contents.

```
>>> x = (1, 2)
>>> x[1] = 3
Traceback (most recent call last):
  File "<stdin>", line 1, in <module>
TypeError: 'tuple' object does not support item assignment
```

A good use for *tuples* is when you are creating data that you want to be "read only." For example, you could create a *tuple* for a material's spring constant and the spring constant units.

```
>>> cork = (20, 'N/m')
```

29
```

In this case, you would always know that *cork[0]* is the value, *cork[1]* is the unit, that neither could ever be changed later in the program, and that *cork[2]* could never be created later. Of course, you could always edit the line to add a third element, but you couldn't do that later on in the program.

A set (*set*) is a type of collection or *list* that contains only unique values.

```
>>> x = set(['x', 'y', 'z', 'x', 'y', 'z'])
>>> x
{'x', 'y', 'z'}
```

The set notation uses curly braces ({ and }), and the above example removed the duplicates *x*, *y*, and *z*. *Sets* can be a quick way to remove any potential duplicates in a *list*:

```
>>> list_of_potential_dups = ['x', 'y', '1', '3', 3, 'z', 'y']
>>> list_of_uniques = list(set(list_of_potential_dups))
```

A dictionary datatype is a "list" of key and value pairs. A dictionary (*dict*) is created with curly braces {}, and its notation is different from a *set*:

```
conversion_dict = {'France': 'French', 'Germany': 'German', 'England': 'English'}
```

The first part of each statement is called the key (*'France'*) and the second part is called the value (*'French'*). The main point of a *dict* is to store a value with a key, and get the value when you give that dictionary a key. Think of it like a normal dictionary: you give the dictionary a word (the key), and it gives you the definition of the word (the value). The notation mimics *str* and *list* slicing, but you need to pass the key, not the index. Dictionaries are unordered, so the order in which you add keys is not necessarily the order in which they are stored. (Read that sentence again. I have been burned by it in the past!)

```
>>> conversion_dict['England']
'English'
```

The most common dictionary functions are *keys()*, which returns a *list*-like object of all of the keys, *values()*, which returns a *list*-like object of all of the values, and *items()*, which provides a *tuple*-like object for each key-value combination. We don't end up using dictionaries in any of the programs in this book, but they are a useful datatype to know about. Note that *{}* represents an empty *dict*, not an empty *set*.

The last datatype is *bool*, short for Boolean. A *bool* can only have one of two values: *True* or *False*. By themselves, *bools* aren't that useful. Their usefulness comes from doing logical comparison operations like equals, does not equal, etc. Below is a list of the logical comparisons in Python:

Operation	Python syntax
Equals	==
Not equals	!=
Less than	<
Less than or equal to	<=

Greater than	>
Greater than or equal to	>=
At least one condition is true	or
Both conditions are true	and

Here are some examples of using logical comparisons:

```
>>> 1 == 2
False
>>> 100000 > 1
True
>>> 5 > 4 and 9 > 8    # could also write (5 > 4) and (9 > 8)
True
>>> 3 > 2 or 1 > -1
True
```

The last things we will cover in this chapter are the *if* statement and the *for* loop. The *if* statement accepts at least one logical condition to check and does something if it is *True*. It is important to note here that Python determines syntax based on whitespace. In other words, indentation level matters in Python. We see this in effect with an *if* statement:

```
>>> x = 5
>>> if x > 3:
>>>     print(x)   # there are four spaces between >>> and p
5
```

The *if* statement is defined with the *if* keyword, a logical operation, a colon, a new line, and an indentation of 4 spaces or 1 tab. Spaces vs tab is another Python community debate; the Spyder IDE will automatically insert 4 spaces if you press the Tab key. In the console, you will have to use the 4 spaces yourself. Anything indented with 4 spaces will be executed if the logical operation evaluates to *True*.

We also learned another function *print()*. *Print()* will print the *str* representation of the variable in its parentheses.

```
>>> print('hello')
hello
```

If you want to print multiple things, you will have to pass them all to the print statement (first method), use the *format()* method (second method), or use f-strings (third method), which were added in Python version 3.6.

```
>>> print('x = ', 4)    # first method
'x = 4'
>>> print('x = {}'.format(4))    # second method
'x = 4'
>>> print(f'x = {x}')    # third method, called f-strings
'x = 4'
```

Since they return a *str*, *format()* or f-strings can be used anywhere you want to ultimately have a *str*. F-strings were added relatively recently, but they have taken the Python world by storm. They are essentially a shortcut way of using the *format()* method. All of the chapters will use *format()* almost exclusively; but, you can use whichever method you prefer.

Now that we have learned about *print()*, let's go back to the *if* statement. The *if* statement can also handle multiple logical operations:

```
>>> x = 5
>>> y = 10
>>> if x > 3 and y <= 10:
>>>     print(x, y)
5, 10
```

If the condition is not *True*, the indented code is simply not run. If you want to catch anything that evaluates to *False*, you can use the *else* keyword. You can also use parentheses to keep your multiple conditions straight.

```
>>> x = 1
>>> y = 10
>>> if (x > 3) and (y <= 10):
        print(x, y)
    else:
        print('x <= 3 or y > 10'.format(x, y))
'x <= 3 or y > 10'
```

You can also evaluate other logical operations using the *elif* statement, which is short for else-if.

```
>>> x = 1
>>> y = 10
>>> if x < 0 and y < 0:
        print('x and y are both negative')
    elif x > 0 and y > 0:
        print('x and y are both positive')
    else:
        print('one is positive and one is negative')
'x and y are both positive'
```

34

Datatypes have inherent "trueness" as well. Essentially, anything that is empty or 0 is *False*, and anything else is *True*.

Datatype	What is False?	What is True?
`int`	0	Anything else
`float`	0.0	Anything else
`str`	`""` (empty `str`)	Any `str` with length > 0
`list`	`[]` (empty `list`)	Any `list` with at least one item
`tuple`	`()` (empty `tuple`)	Any `tuple` with at least one item
`set`	`set()` (empty `set`)	Any `set` with at least one item
`dict`	`{}` (empty `dict`)	Any `dict` with at least one key and value
`bool`	`False`	`True`

This means that you can abbreviate `if int != 0` to `if int`, for example. Be careful as `bool("False")` evaluates to *True* because "False" is a *str* of length > 0.

The last thing we will review in this chapter is the *for* loop. The *for* loop takes an iterable object like a *list* and gives you each element within that object, one at a time.

```
numbers = [1, 2, 3, 4, 5]
for number in numbers:
    print(number)
```

Remember how an advantage of Python is that it is very readable? Just by looking at that snippet of code, it is somewhat obvious what the code is doing. It takes each number from *numbers* and prints it. It also does not matter what word you use in place of *number*; the Python interpreter understands that you are addressing each element of the *list*. We will use the *for* loop in the next section. In the console, practice using the *for* loop to get comfortable with it. What happens if you change the variable *number*?

One last piece of advice: if you want something to exist after a loop, it needs to exist before the loop. In other words, don't create variables inside of *for* loops if you want to use them again. Given a list of numbers 1-10, this is one way you could use a *for* loop to keep only the even numbers:

```
end_list = []
numbers = [1, 2, 3, 4, 5, 6, 7, 8, 9, 10]
for number in numbers:
    if number % 2 == 0:
        end_list.append(number)
print(end_list)
>>> [2, 4, 6, 8, 10]
```

The *for* loop has a cousin called the *while* loop. The *while* loop repeatedly runs a block of code until the condition is no longer *True*.

```
>>> i = 0
>>> while i < 4:
        print(i)
        i += 1
0
1
2
```

If we didn't add the *i += 1* part, the *while* loop would run forever since that condition is always *True*. At the end of the loop, *i = 4*. We won't end up using *while* loops in this book, but they are a good control sequence complement to the *for* loop.

Here is a review of the datatypes we discussed:

Datatype	Explanation
`int`	Integer. A whole number.
`float`	Floating point number. A number with at least one decimal place.
`str`	String. Any type of text representation.
`list`	A collection of things that is ordered and changeable.
`tuple`	A collection of things that is ordered and unchangeable.
`set`	A collection of things that is unordered and contains only unique values.
`dict`	A collection that is unordered and unindexed. It has keys and values.
`bool`	Boolean, meaning True or False.

In the next section, we will put all of this information to use with our first program: FizzBuzz.

Resources

[1] https://docs.scipy.org/doc/numpy/user/numpy-for-matlab-users.html

[2] https://github.com/alexkenan/pymae

Chapter 1.5: FizzBuzz

Introduction

Each chapter of this book will have an introduction to the problem or program we are trying to create, an import section for any libraries we need to import, the body of the program as it is developed, a conclusion or summary, and the full code at the end. This format should provide enough context to understand the general idea of the program and should allow you to pick through what you need to know to get started. Links to any third-party library official website will also be included, which usually has a documentation section that will have an introduction to the library and show examples. This documentation is very helpful in learning more about the specific library, or for troubleshooting pieces of code that will not cooperate.

In software development, "FizzBuzz" is a common first-level interview question that weeds out anyone who cannot program in the desired language. The question generally goes something like this:

Write a program that prints out the numbers 1 to 100. If the number is a multiple of 3, print *Fizz* instead of the number. If the number is a multiple of 5, print *Buzz* instead of the number. If the number is a multiple of both 3 and 5, print *FizzBuzz* instead of the number.

We will use the crash course in Chapter 1 to complete FizzBuzz.

Imports

We won't be using any other libraries, so there are no imports. Ordinarily, imports will go at the very beginning of your program to be used later.

Body

In Spyder, create a new file and save it as *fizzbuzz.py*. This file will have the completed FizzBuzz Python program. *range()* is a built-in Python function that accepts at least one argument and two optional ones: the optional beginning of the desired range (default starting point is 0 if none is specified), the end of the desired range that will not be included, and an optional *int*, which specifies how much to step or increment the range (default is 1). *range()* only accepts integers. *range(1, 4)* will provide 1, 2, and 3, but not 4. *range(1, 10, 2)* will provide 1, 3, 5, 7, and 9. We will also need to make use of the modulus (%) operator. The line *x % y* returns the remainder if *x* were to be divided by *y*. A return of 0 means that *x* is evenly divisible by *y*.

We can use *range()* and the *for* loop to iterate through the numbers 1 to 100, use a few *if* statements to see if that number is divisible by 3, 5, or 3 and 5, and print out the number, Fizz, Buzz, or FizzBuzz as appropriate. Type the following code into the text editor area.

```
for number in range(1, 100):
    if number % 3 == 0:
        print('{} Fizz'.format(number))
    elif number % 5 == 0:
        print('{} Buzz'.format(number))
    elif (number % 3 == 0) and (number % 5 == 0):
        print('{} FizzBuzz'.format(number))
    else:
        print(number)
```

Let's see the results from numbers 10 through 15 since there should be a Buzz (10), a Fizz (12), and a FizzBuzz (15):

```
10 Buzz
11
12 Fizz
13
14
15 Fizz
```

We got a *Fizz* but not a *FizzBuzz* for 15. Why is that? Remember, the indented code block gets executed if the logical operation is true. Since, by definition, multiples of 3 **and** 5 will also be a multiple of 3 **or** a multiple of 5, the FizzBuzz part will never execute, because either the multiple of 3 part will be *True*, or the multiple of 5 part will be *True*, and the multiple of 3 and 5 part will never get reached. We can fix this problem by making the multiple of 3 and 5 part be the first *if* statement, and then multiple of 3 or multiple of 5 *elif* statements come later.

```
for number in range(1, 100):
    if (number % 3 == 0) and (number % 5 == 0):
        print('{} FizzBuzz'.format(number))
```

```
    elif number % 5 == 0:
        print('{} Buzz'.format(number))
    elif number % 3 == 0:
        print('{} Fizz'.format(number))
    else:
        print(number)
```

There's still one issue with the code; let's take a look at 95-100. See if you can spot the issue.

```
95 Buzz
96 Fizz
97
98
99 Fizz
```

We're missing 100! *range()* doesn't give you the end number, so *range(1, 100)* will only give us 1 through 99. We need to change it to *range(1, 101)* to get 100.

```
for number in range(1, 101):
    if (number % 3 == 0) and (number % 5 == 0):
        print('{} FizzBuzz'.format(number))
    elif number % 5 == 0:
        print('{} Buzz'.format(number))
    elif number % 3 == 0:
        print('{} Fizz'.format(number))
    else:
        print(number)
```

If we want to be faithful to the exact way the problem was worded, we can get rid of the *format()* functions for any number that gets a *Fizz*, *Buzz*, or *FizzBuzz*.

```
for number in range(1, 101):
    if (number % 3 == 0) and (number % 5 == 0):
        print('FizzBuzz')
    elif number % 5 == 0:
        print('Buzz')
    elif number % 3 == 0:
        print('Fizz')
    else:
        print(number)
```

Congratulations, you've written your first Python program!

Conclusion

FizzBuzz is a common interview technique to get a sniff-test of the candidate: can this candidate actually write code in the desired language? Despite being such a relatively straight-forward task, there are many different ways to accomplish this end result. For example, we could have used a *while* loop instead of a *for* loop. There are also lots of different variations of the FizzBuzz task, but they all answer the same questions: does the person have an understanding of *if* statements, control sequences (*for* and *while* loops), integer division (or modulus operator), and the logical sequence of commands? FizzBuzz is also a good way to show the basic syntax and structure of a programming language. Now that we have FizzBuzz down, we can move onto more complicated programming structures.

Full Code

```python
for number in range(1, 101):
    if (number % 3 == 0) and (number % 5 == 0):
        print('FizzBuzz')
    elif number % 5 == 0:
        print('Buzz')
    elif number % 3 == 0:
        print('Fizz')
    else:
        print(number)
```

View this code on GitHub at https://github.com/alexkenan/pymae/

Chapter 2: Graphing Thrust Available and Thrust Required

Introduction

The first example that we will use to show off Python's functionality is graphing thrust required and thrust available for an Airbus A321 at sea level, 10,000 feet, and 35,000 feet (often stylized as FL350 for Flight Level 350, meaning 35,000 feet). This example allows us to demonstrate the plotting capabilities of Matplotlib, the graphing library, and Python's ability to accept functional programming inputs like writing equations and iterating over *lists*. Explaining the relationship between thrust available, thrust required, and drag is beyond the scope of this book; however, we will make use of the fact that thrust required equals drag at the steady, level, unaccelerated cruise condition to calculate thrust required by calculating drag. We will also make the assumption that thrust available at different altitude scales with the ratio of the density of air, so sea level thrust is the nominal thrust of both engines, and thrust output at 35,000 feet is the sea level thrust times the ratio of the density of air at 35,000 feet over the density of air at sea level. As an equation:

$$Thrust_{FL350} = Thrust_{SL} * \frac{\rho_{FL350}}{\rho_{SL}}$$

It's not exactly correct, but it is a decent assumption for this example.

The program will iterate through a range of velocities to calculate the thrust required at each velocity, then plot thrust as a function of velocity for sea level. The program will then repeat the calculation for 10,000 ft and 35,000 ft. We will use the Matplotlib library for the graphing, and since we will need to use pi, we'll import pi from Numpy to use it.

The formatting for this chapter may be off for any lines with *plt.plot()* or long individual lines of code. If you run into any errors, take a look at the programs at the GitHub site to ensure the lines are correct. The formatting looked great in Microsoft Word, but the conversion to ebook format can break lines in unexpected places.

Imports

```
from numpy import pi
import matplotlib.pyplot as plt
```

These two lines go at the top of the new program in the text editor area. Since we don't need the entire Numpy library, we can import specific parts of it with the *from x import y* syntax. Calling the Matplotlib object *plt* is a common convention, but Python supports any sort of naming convention as long as you are consistent. Both libraries are included with Anaconda and Spyder, so nothing should error out. If it does, open up a CMD Prompt or Terminal and type *pip install numpy matplotlib* which should download the libraries.

Body

The first thing we want to do is write two functions that we will use as equations to convert from knots to feet per second and to calculate thrust required from the given parameters. Let's use the equation $y = x^2$ to explain what a function is in Python. We can write a function that takes x, turns it into x^2, and returns x^2 back to what originally called the function. This characteristic means that we don't have to keep pasting in the equation every time we want to use it; we can just invoke the function to calculate it for us. It is easy enough to write x^2 each time, but it would become more annoying to keep pasting a more complicated equation. The other benefit is that we can edit the function's calculation in one place, and those changes will be reflected immediately every time we call that equation. If we suddenly wanted the function to calculate x^3 instead of x^2, we wouldn't have to go hunt for every x^2 in the code to change. You create a function in Python by starting with the *def* keyword. For example:

```
def squared(x):
    return x*x     # or x**2. Remember x^2 won't work!
```

That function would accept any number and *return* its square. You call the function by using its name and passing the argument *x*, which can be any variable name:

```
>>> a = 5
>>> y = squared(a)
>>> y
25
```

You need to define the *squared* function before you try calling it; therefore, most functions will go at the top of the program after the imports so that they can be called later in the program. This next function will take *x* and print its square to console, but it doesn't return a value for *x*:

```
def squared2(x):
    print(x*x)
```

If you tried the same x = 5, y = squared2(x) as above, 25 will be printed to the console, but *y* will be of type *None*, not 25, because we did not return anything.

Starting in Python 3, you can use type hinting to suggest what datatype *x* is, and what the function will return. You can also create docstrings, which are blurbs that describe what the function does, what parameters it accepts, and what it returns. This is like writing a big comment explaining what your function does, and it is always beneficial (and sometimes tedious or annoying) to write one for every function. Here is the original *squared* function with type hinting and a docstring:

```
def squared(x: float) -> float:
    """
    Takes x as a float and returns its square
    :param x: float
    :return x**2
    """
    return x*x
```

The type hinting says that the *squared* function is expecting a *float* and will return a *float*. Now, in your program, if you type *squared(* with just the open parenthesis, Spyder should have a popup textbox that reminds you what the argument(s) should be. Now that we have learned about functions, let's write two.

The first will convert knots (kts) to feet per second (fps). Fps is the base unit for velocity in US Imperial units, but most aircraft use knots for velocity. We will show the x-axis of the graph in knots and dynamically calculate the thrust in converting knots to fps.

```
def knots_to_ftpersec(speed):
    # speed is the user input and is in knots
    return speed * 1.68781
```

The second function takes density, velocity, wing area, zero-lift drag coefficient Cd0, drag constant K, and aircraft weight, and returns the thrust required in lbs. The equation itself is derived from the fact that drag equals thrust at altitude and unbundles the drag equation into its separate parts. Here is the equation:

$$C_L = \frac{2 * Weight}{\rho_\infty * V_\infty^2 * S}$$

In Python:

```python
def thrust_required(rho_inf, v_inf, s, cd0_, k, w):
    cl = 2 * w / (rho_inf * v_inf ** 2 * s)
    return 0.5 * rho_inf * v_inf ** 2 * s * (cd0_ + k * cl ** 2)
```

Another reminder: the carat operator ^ does not perform exponentiation in Python; it is a logical operator for "bitwise exclusive or." That is computer science talk for "will not compute what you think it will compute." The double asterisk ** is what performs exponentiation, so be careful to use the correct operator.

Next, we will define the inputs that we are using. The various Airbus A321 aircraft parameters are estimated from different sources around the internet. They aren't exactly right, but they are close enough for our purposes.

```python
weight = 200000      # lb
wing_area = 1318     # ft^2
wing_span = 117.416666667   # ft
cd0 = 0.0185
thrust = 66000       # lb of thrust total
aspect_ratio = wing_span**2 / wing_area
e = 0.92
K = 1 / (pi * e * aspect_ratio)
rho_sl = 23.77E-4    # in slugs/ft^3
```

Now, we will create the velocities to sweep between. The beginning and end ranges were guess-and-checked to make sure that both thrust required and thrust available intercepts would be displayed. We will use *range()* to generate a *list* of numbers:

```
x_vals_sl = []
for i in range(80, 750, 10):
    x_vals_sl.append(i)
```

That *for* loop says for each value in *range()*, add it to the *list* that we defined above. Remember, if we modify something in a *for* loop and want it to exist afterwards, we need to define it beforehand.

There is shorthand for the above statement that we can invoke using a list comprehension. A list comprehension essentially flattens the *for* loop to be on one line, and you avoid having to define the *list* before the *for* loop:

```
x_vals_sl = [i for i in range(80, 750, 10)]
```

We will do a similar thing to create the thrust values.

```
tr_sl = []   # list of thrust required values at sea level
for airspeed in x_vals_sl:
    airspeed_fps = knots_to_ftpersec(airspeed)
    thrust_calculated = thrust_required(rho_sl, airspeed_fps,
                                        wing_area, cd0, K, weight)
    tr_sl.append(thrust_calculated)
```

Indentation level matters in Python, but after an open parenthesis (, you can use whatever indentation level you want. Due to margin formatting for this book, code has to be broken up to make it fit onto the page. In Spyder, the line that starts with *thrust_calculated* = would be on one line. There is a Python convention to break up lines that are more than 80 characters long. If you do this, the convention is to continue the indentation under the most recent (. So, the "correct" way to break the line is:

```
thrust_calculated = thrust_required(rho_sl, airspeed_fps,
                                    wing_area, cd0, K, weight)
```

We try to follow this convention, but the formatting in the book may look different to fit inside the page margins.

The *for* loop could be flattened to one line inside of the loop if we dynamically calculate the fps value and thrust value:

```
tr_sl = []
for airspeed in x_vals_sl:
    tr_sl.append(thrust_required(rho_sl, knots_to_ftpersec(x),
                                 wing_area, cd0, K, weight))
```

And, of course, by writing the *for* loop like this, we can make it one line with another list comprehension:

```
tr_sl = [thrust_required(rho_sl, knots_to_ftpersec(x), wing_area,
                         cd0, K, weight) for x in x_vals_sl]
```

All three statements are equivalent. List comprehensions can be difficult to read at times (especially the second one we wrote), so I tend to avoid their use. They are handy shortcuts though, so you should at least know about them.

Now, we have the thrust required values (y-values) and the velocities (x-values), so we'll let Matplotlib plot them. The syntax is fairly easy:

```
plt.plot(x_vals_sl, tr_sl, 'k-', label=r"$T_R$ at Sea Level")
```

With 'k-', we are telling Matplotlib to plot a black, solid line. The *Label* argument is the label that will go in the legend we create. The T_R part is LaTeX, pronounced lah-teck, which provides equation-editor functionality to create the T_R subscript. Here are the different line styles and color options in Matplotlib:

Linestyle	Description
'-' or 'solid'	solid line
'--' or 'dashed'	dashed line
'-.' or 'dashdot'	dash-dotted line
':' or 'dotted'	dotted line
'None' or ' ' or ''	draw nothing

Character	Color
'b'	blue
'g'	green
'r'	red
'c'	cyan
'm'	magenta
'y'	yellow
'k'	black

Matplotlib has several different cheatsheets and handouts available at its GitHub site [1]. If you want to customize your color options, you can pass a specific color argument and use HTML-style hex color codes. This advanced usage is shown in Chapter 5.

We'll assume a 70% throttle setting, and we'll use another list comprehension to build a line parallel to the x-axis that will represent thrust available, which does not change with velocity.

```
TA_sl = 0.7 * thrust
y_coords_sl = [TA_sl for _ in x_vals_sl]
```

Notice how we use the underscore _ to represent that we don't care what the actual value we're getting out of *x_vals_sl* is; we just want to make a *List* of all *TA_sl* values. Now we can plot the thrust available line:

```
plt.plot(x_vals_sl, y_coords_sl, 'k--',
         label='$T_A$ at Sea Level ({:,.0f} lb)'.format(TA_sl))
```

Lastly for plotting, we'll make a vertical line at 180 knots between the two curves to show the takeoff velocity.

```
takeoff_vel_sl = [180, 180]
cruise_velocity_fl350_values = [11000, 46500]

plt.plot(takeoff_vel_sl, cruise_velocity_fl350_values, 'b-.',
         label="Takeoff Velocity ({} knots)".format(takeoff_vel_sl[0]))
```

As shorthand, we format the takeoff velocity as the first element in *takeoff_vel_sl* so we don't have to repeatedly type the value.

The only remaining steps are to customize the plot. We need to set x- and y-axis limits, label both axes, place the legend, give the plot a title, and, finally, show the plot.

```
plt.ylim(5000, 50000)
plt.xlim(50, 1550)
plt.ylabel('Thrust (lb)')
plt.xlabel('Velocity (kts)')
plt.title('Thrust Required & Thrust Available Curves')
plt.legend(loc='lower right')
plt.show()
```

Spyder has a new setting where it creates a new *Plots* tab in the variable explorer area. This is where your plot will go by default. To also see the plot in-line in the Notebook, go to the menu in the *Plots* pane and un-check "Mute inline plotting."

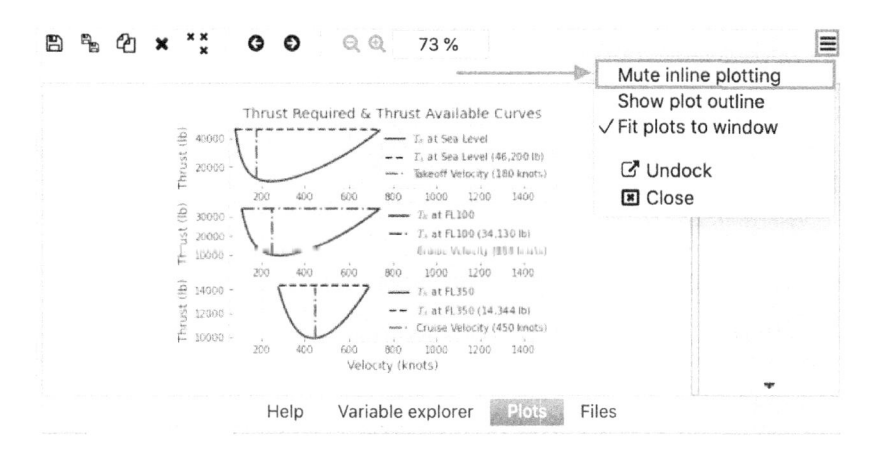

This is what your plot should look like:

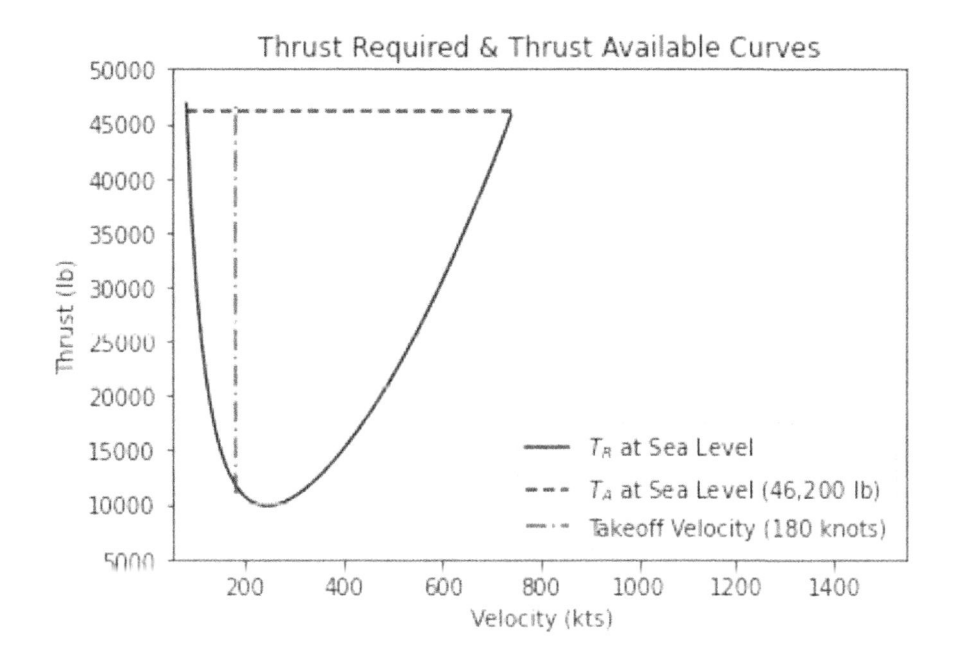

We could probably adjust the x-axis a little more to remove some of the whitespace, but it looks good, and it will help to have the space later. Your plot may look slightly different based on different computers and formatting, but the overall shape should look the same.

We want to create three subplots to be shown that represent each altitude, so we need to tell Matplotlib that we want to make subplots. After the inputs but before the sea level code, add:

```
plt.subplot(3, 1, 1)
```

We are telling Matplotlib that we intend to have 3 rows, 1 column, and this section is the first subplot we are creating. The graph, when complete, should look something like this:

	Column 1
Row 1	Sea Level graph
Row 2	10,000 ft graph
Row 3	FL350 graph

For the 10,000 ft graph, the density, takeoff velocity (which we will now call cruise velocity), velocities to iterate through, thrust available, and thrust required all change. Here is how we will reflect the changes:

```
plt.subplot(3, 1, 2)
rho_fl100 = 17.56E-4    # in slugs/ft^3
x_vals_fl100 = [i for i in range(110, 745, 10)]
tr_fl100 = [thrust_required(rho_fl100, knots_to_ftpersec(x),
```

```
                    wing_area, cd0, K, weight) for x in x_vals_fl100]
TA_fl100 = 0.7 * thrust * rho_fl100 / rho_sl
cruise_velocity = [250, 250]
cruise_velocity_fl100_values = [10000, 34130]
y_coords_fl100 = [TA_fl100 for _ in x_vals_fl100]
```

We will use 250 knots as the cruise velocity since that is the "speed limit" for aircraft below 10,000 ft. Calling 10,000 feet "FL100" is technically incorrect since, in the US, flight levels start at 18,000 ft. But, it is convenient shorthand, so we'll use it. We can plot the same three sets of coordinates, and do the same labeling and axis-adjusting:

```
plt.plot(x_vals_fl100, tr_fl100, 'k-', label=r"$T_R$ at FL100")

plt.plot(x_vals_fl100, y_coords_fl100, 'k--',
         label="$T_A$ at FL100 ({:,.0f} lb)".format(TA_fl100))

plt.plot(cruise_velocity, cruise_velocity_fl100_values, 'b-.',
         label="Cruise Velocity ({} knots)".format(cruise_velocity[0]))

plt.ylabel('Thrust (lbs)')
plt.ylim(5000, 38000)
plt.xlim(50, 1550)
plt.legend(loc='lower right')
plt.show()
```

The *{:,.0f}* is ensuring that we get a comma-separated number that is rounded to 0 decimal places. Make sure that you delete the *plt.show()* from the sea level section; otherwise, Matplotlib will show you two charts, and they won't be in the same window.

Thrust Required & Thrust Available Curves

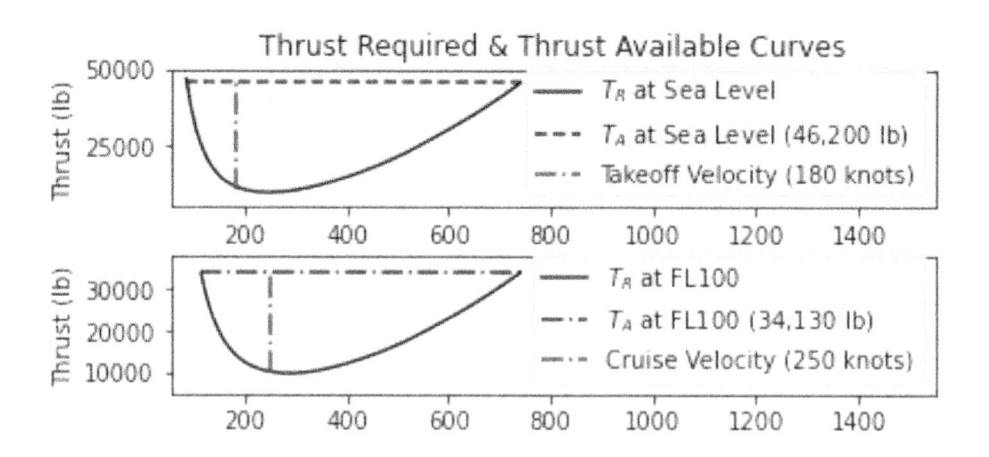

- T_R at Sea Level
- T_A at Sea Level (46,200 lb)
- Takeoff Velocity (180 knots)
- T_R at FL100
- T_A at FL100 (34,130 lb)
- Cruise Velocity (250 knots)

If your subplots are overlapping each other, add `plt.subplots_adjust(hspace=0.30)` before `plt.show()` to create some vertical space between the subplots. Tweak the number (`0.30`) to get padding that looks just right.

We'll mimic the same changes that we made going from sea level to 10,000 ft when we go from 10,000 ft to FL350. Let's skip the individual sections and show the new section below. Remember, you only want one `plt.show()` at the end of your program. Anything that referenced `fl100` gets changed to `fl350`.

```
plt.subplot(3, 1, 3)
rho_fl350 = 7.38E-4   # in slugs/ft^3
cruise_velocity_fl350 = [450, 450]
cruise_velocity_fl350_values = [9900, 14200]
TA_fl350 = 0.7 * thrust * rho_fl350 / rho_sl
x_vals_fl350 = [i for i in range(280, 695, 10)]
tr_fl350 = [thrust_required(rho_fl350, knots_to_ftpersec(x),
            wing_area, cd0, K, weight) for x in x_vals_fl350]

plt.plot(x_vals_fl350, tr_fl350, 'k-', label=r"$T_R$ at FL350")
y_coords_fl350 = [TA_fl350 for _ in x_vals_fl350]
plt.plot(x_vals_fl350, y_coords_fl350, 'k--',
         label=r"$T_A$ at FL350 ({:,.0f} lb)".format(TA_fl350))
```

```
plt.plot(cruise_velocity_fl350, cruise_velocity_fl350_values, 'b-.',
        label="Cruise Velocity ({} knots)".format(
                                    cruise_velocity_fl350[0]))

plt.xlabel('Velocity (knots)')
plt.ylabel('Thrust (lb)')
plt.xlim(50, 1550)
plt.legend(loc='lower right')
plt.show()
```

Here is the creation!

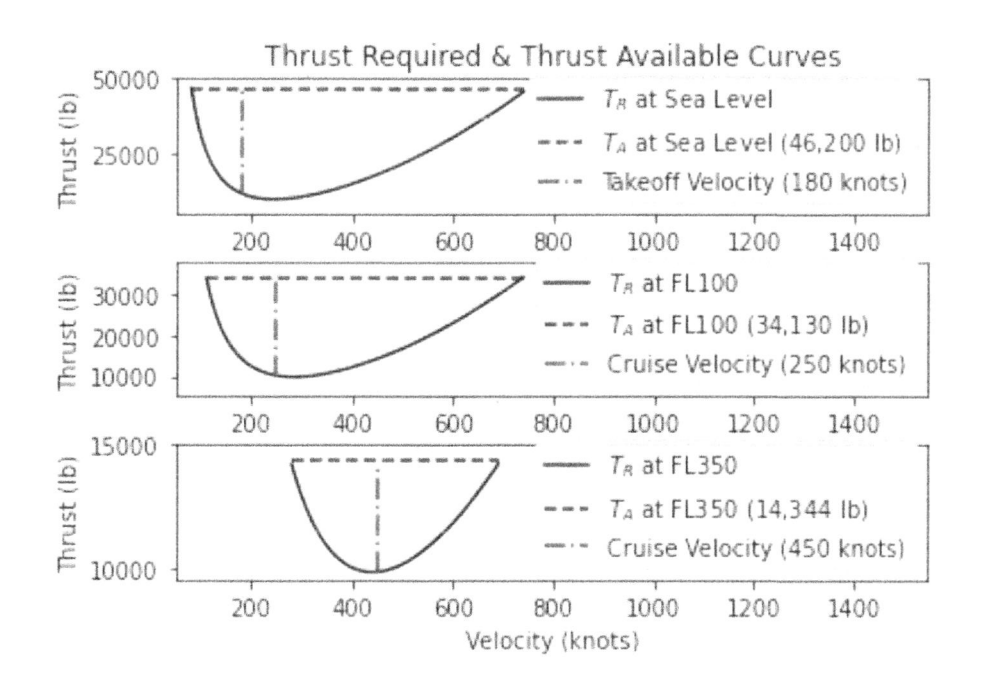

That's a nice looking graph! We had to squish the axes between the different subplots to get them to look approximately similar; the points could have been graphed on the same plot which would have provided a better representation of just how much thrust decays with altitude. Notice how we end up copy-and-pasting a lot of the same code over and over again? In particular, the section where we create the three sets of coordinates is just copy-and-pasted from the last section, and only the variable names and constants are changed. This program could be made a lot more Pythonic by writing a *plot_values()* function that accepts the three sets of values and plots them. As it is, the program is a good introduction to functional programming and using Matplotlib to make graphs.

Summary

We were able to learn about functions in Python, use them to convert knots to fps and calculate thrust required, and graph the results by passing the coordinates to Matplotlib. We also discussed how we could make the program a little more Pythonic and concise by using the principle of Don't Repeat Yourself (DRY) to be able to write functions so that we don't have to repeat code. We still made a great program that graphically shows how thrust available decreases with altitude, and how the thrust curve's shape changes with the drag characteristic at altitude. Once you are comfortable with writing this chapter's program, try making the program display the same graphs but by plotting all values via one *plot_values()* function that you define. There is an example of how to plot all of the points on one plot at the GitHub site in the Chapter 2 folder.

Full Code

```python
from numpy import pi
import matplotlib.pyplot as plt


# Functions
def knots_to_ftpersec(speed):
    return speed * 1.68781

def thrust_required(rho_inf, v_inf, s, cd0 , k, w):
    cl = 2 * w / (rho_inf * v_inf ** 2 * s)
    return 0.5 * rho_inf * v_inf ** 2 * s * (cd0_ + k * cl ** 2)

# Inputs

weight = 200000   # lb
wing_area = 1318   # ft^2
wing_span = 117.416666667   # ft
cd0 = 0.0185
thrust = 66000   # lb of thrust total
aspect_ratio = wing_span**2 / wing_area
e = 0.92
K = 1 / (pi * e * aspect_ratio)
rho_sl = 23.77E-4   # in slugs/ft^3
x_vals_sl = [i for i in range(80, 750, 10)]
tr_sl = [thrust_required(rho_sl, knots_to_ftpersec(x), wing_area,
                        cd0, K, weight) for x in x_vals_sl]


# Sea Level
plt.subplot(3, 1, 1)
plt.plot(x_vals_sl, tr_sl, 'k-', label=r"$T_R$ at Sea Level")
TA_sl = 0.7 * thrust
y_coords_sl = [TA_sl for _ in x_vals_sl]
takeoff_vel_sl = [100, 100]
cruise_velocity_sl_values = [11000, 46500]
plt.plot(x_vals_sl, y_coords_sl, 'k--',
         label='$T_A$ at Sea Level ({:,.0f} lb)'.format(TA_sl))
```

```python
plt.plot(takeoff_vel_sl, cruise_velocity_sl_values, 'b-.',
         label="Takeoff Velocity ({} knots)".format(
                                    takeoff_vel_sl[0]))
plt.ylim(5000, 50000)
plt.xlim(50, 1550)
plt.ylabel('Thrust (lb)')
#plt.xlabel('Velocity (kts)')
plt.title('Thrust Required & Thrust Available Curves')
plt.legend(loc='lower right')

# FL100
plt.subplot(3, 1, 2)
rho_fl100 = 17.56E-4  # in slugs/ft^3
x_vals_fl100 = [i for i in range(110, 745, 10)]
tr_fl100 = [thrust_required(rho_fl100, knots_to_ftpersec(x),
                  wing_area, cd0, K, weight) for x in x_vals_fl100]
TA_fl100 = 0.7 * thrust * rho_fl100 / rho_sl
cruise_velocity = [250, 250]
cruise_velocity_fl100_values = [10000, 34130]
y_coords_fl100 = [TA_fl100 for _ in x_vals_fl100]

plt.plot(x_vals_fl100, tr_fl100, 'k-', label=r"$T_R$ at FL100")
plt.plot(x_vals_fl100, y_coords_fl100, 'k--',
         label="$T_A$ at FL100 ({:,.0f} lb)".format(TA_fl100))
plt.plot(cruise_velocity, cruise_velocity_fl100_values, 'b-.',
         label="Cruise Velocity ({} knots)".format(
                                    cruise_velocity[0]))
plt.ylabel('Thrust (lb)')
plt.ylim(5000, 38000)
plt.xlim(50, 1550)
plt.legend(loc='lower right')

# FL350
plt.subplot(3, 1, 3)
rho_fl350 = 7.38E-4  # in slugs/ft^3
cruise_velocity_fl350 = [450, 450]
cruise_velocity_fl350_values = [9900, 14200]

x_vals_fl350 = [i for i in range(280, 695, 10)]
tr_fl350 = [thrust_required(rho_fl350, (knots_to_ftpersec(x)),
```

```
                    wing_area, cd0, K, weight) for x in x_vals_fl350]
TA_fl350 = 0.7 * thrust * rho_fl350 / rho_sl
y_coords_fl350 = [TA_fl350 for _ in x_vals_fl350]
plt.plot(x_vals_fl350, tr_fl350, 'k-', label=r"$T_R$ at FL350")
plt.plot(x_vals_fl350, y_coords_fl350, 'k--',
         label=r"$T_A$ at FL350 ({:,.0f} lb)".format(TA_fl350))
plt.plot(cruise_velocity_fl350, cruise_velocity_fl350_values, 'b-.',
         label="Cruise Velocity ({} knots)".format(
                                        cruise_velocity_fl350[0]))

plt.xlabel('Velocity (knots)')
plt.ylabel('Thrust (lb)')
plt.xlim(50, 1550)
plt.ylim(9600, 15000)
plt.legend(loc='lower right')
# might need this next line
#plt.subplots_adjust(hspace=0.30)
plt.show()
```

View this code on GitHub at https://github.com/alexkenan/pymae/

References

[1] https://github.com/matplotlib/cheatsheets

Chapter 3: Graphing Dynamic Pressure during a Rocket Launch

Introduction

If you have ever watched a broadcast of a rocket launch, about 30 seconds after liftoff, the announcer indicates that the rocket is lowering the throttles for "Max Q" or maximum dynamic pressure. Max Q represents the maximum pressure exerted on the rocket by the air as the rocket moves through the atmosphere. We will use Matplotlib to graphically represent pressure as a function of time to understand where Max Q occurs and what the total pressure is at Max Q for different average accelerations. Dynamic pressure is a function of the density of air (which changes with altitude and temperature) and velocity (which will change with time as the rocket speeds up).

The program will calculate dynamic pressure as a function of time, find the maximum pressure and the time at which it occurs, plot dynamic pressure vs time, and identify and annotate Max Q with its index and value. Being able to annotate the graph with arrows and textboxes is another powerful Matplotlib feature that we will learn about. It will show how changing launch duration can have dramatic effects on the maximum dynamic pressure.

Science & Physics Discussion

This section explains the assumptions made to ultimately calculate dynamic pressure as a function of time. This requires a brief discussion about physics. If you would rather get right to the programming, skip to the next section.

The equation for dynamic pressure q is 0.5 * rho * V^2 where rho is the density of air and V is velocity. We want to build two equations so that we have density and velocity as functions of time (so that we can calculate dynamic pressure as 0.5 * rho(t) * $V(t)^2$).

Velocity. The rocket needs to go from 0 mph to about 18,000 mph, and the average rocket launch lasts about 8.5 minutes to go from launchpad to "space" [1]. Thus, we will use the average acceleration of 18,000 mph / 8.5 minutes = 51.76 ft/s^2. We will also assume that this acceleration is constant (e.g. the rocket is constantly accelerating at 51.76 ft/s^2 from 0 mph to 18,000 mph), which allows us to use simple physics equations. This is an incorrect assumption, but we will use it to start, then we can refine the average acceleration as needed. For constant acceleration, velocity$_{final}$ = velocity$_{initial}$ + acceleration * time. Since the rocket is starting motionless on the launchpad, velocity$_{initial}$= 0. We now have a Python function that will calculate velocity based off of acceleration and time. It will look something like this:

```
instant_velocity = velocity(acceleration, time) = acceleration * time
```

Density. The density of air scales with temperature which decreases with higher altitude. For density, we will need to separately create an equation for altitude from velocity, use that altitude to get air temperature at that altitude, then use that air temperature to get density. We will use another constant acceleration toolbox equation to get altitude as a function of time. The basic equation is:

$$\text{altitude} = \text{velocity}_{\text{initial}} * \text{time} + 0.5 * \text{acceleration} * \text{time}^2$$

Velocity$_{\text{initial}}$ is 0 again, so the equation becomes altitude = 0.5 * acceleration * time^2. In Python pseudocode:

```
Altitude = time_to_altitude(acceleration, time) = 0.5 * acceleration * time ** 2
```

One last reminder: in order to calculate exponents, use the ** operator.

NASA has a basic altitude-temperature model [2] that shows the relationship between altitude, temperature, and density. We can build this model into a *density()* equation that takes altitude as an input, internally calculates temperature, calculates density, then returns density. In Python pseudocode:

```
density = density(altitude) =
temperature_to_density(altitude_to_temperature(altitude))
```

This is a nested argument that we discussed in the Introduction; the Python interpreter will evaluate the line from inside out and dynamically calculate each function. In other words, it will evaluate *altitude_to_temperature(altitude)* to calculate a temperature from altitude, then evaluate *temperature_to_density(temperature)* from the result to calculate density from temperature. Now, we should be able to calculate dynamic pressure since we have equations for density and velocity written as functions of time. One last time, in Python pseudocode:

```
dynamic_pressure(density(altitude), velocity(acceleration, time)) =
0.5 * density(time_to_altitude(acceleration, time)) *
velocity(acceleration, time)**2
```

Now, back to the regularly scheduled Python programming.

Imports

```
import matplotlib.pyplot as plt
import numpy as np
```

Numpy will be used below to generate ranges of values for this tutorial in 0.5 second increments and to perform exponential calculations. The built-in *range()* function used in FizzBuzz could very easily be used in Numpy's place to generate a range of numbers. The built-in *range()* only accepts integer increments, which means you would lose the ability to increment by numbers less than 1. So, let's use Numpy's *arange()* which accepts decimal increments.

67

Body

We will fully write the three functions we developed in the Science & Physics Discussion section below:

```python
def density(height: float) -> float:
    if height < 36152.0:
        temp = 59 - 0.00356 * height
        p = 2116 * ((temp + 459.7)/518.6)**5.256
    elif 36152 <= height < 82345:
        temp = -70
        p = 473.1*np.exp(1.73 - 0.000048*height)
    else:
        temp = -205.05 + 0.00164 * height
        p = 51.97*((temp + 459.7)/389.98)**-11.388

    rho = p/(1718*(temp+459.7))
    return rho


def velocity(time: float, acceleration: float) -> float:
    return acceleration*time


def altitude(time: float, acceleration: float) -> float:
    return 0.5*acceleration*time**2
```

Now, we get to the setup part of the code to get ready to perform the calculations. We will declare what style template we want the plot to look like, create an empty *list* of *x_values* (which will become time in seconds) and *y_values* (which will become dynamic pressure values in pounds per square foot or psf). We will use *np.arange()* to iterate from 0 seconds to 550 seconds in 0.5 second increments for a total of 1,100 values to compute.

```
plt.style.use('bmh')
y_values = []
x_values = np.arange(0.0, 550.0, 0.5)
```

Next comes the meat and potatoes of the code. We will iterate through each x value within *x_values*, calculate velocity, altitude, density, and ultimately dynamic pressure, and then store the dynamic pressure in *y_values*.

```
for elapsed_time in x_values:
    accel = 51.76 # ft/s^2
    alt = altitude(elapsed_time, accel)
    # Dynamic pressure q = 0.5 * density * velocity^2
    q = 0.5 * density(alt) * velocity(elapsed_time, accel) ** 2
    y_values.append(q)
```

The points are ready to plot. All we have to do is call the *plot()* function and give the same few customizable parameters as in the previous chapter:

```
plt.plot(x_values, y_values, 'b-',
         label=r"a = 51.76 $\frac{ft}{s^2}$")
```

If you need a refresher on linestyle and linecolor options, go back to Chapter 2.

We also want to be able to mark on the graph what the maximum pressure is and where it occurs; by marking this point, we can say at what time Max Q occurs and what value the maximum dynamic pressure reaches. We will use the built-in *max()* function to determine the greatest number in *y_values*, and we will use the *index()* function to determine where that value is in the *x_values*. Then, we can use that index to find out what elapsed time the highest value corresponds to. It is important to note that the index is the location in the list of where the Max Q is in *y_values* and the time at which it occurs in *x_values*; it is not the time at which Max Q occurs, nor is it the value of Max Q.

```
max_val = max(y_values)
ind = y_values.index(max_val)
```

So, *x_values[ind]* is the time at which Max Q occurs, and *y_values[ind]* is the value of Max Q. We will use *pyplot.annotate()* function to add an arrow pointing to the max value and to add a label with the value and time at which Max Q occurs. In Python:

```
plt.annotate('{:.0f} psf @ {} s'.format(max_val, x_values[ind]),
            xy=(x_values[ind] + 2, max_val),
            xytext=(x_values[ind] + 15, max_val + 15),
            arrowprops=dict(facecolor='blue', shrink=0.05),
            )
```

{:.0f} means that the annotation will have 0 decimal places. The *plt.annotate()* argument accepts the label text (which is populated with Max Q and the time at which it occurs), x and y values to start the arrow, x and y values to start the label text, and a dictionary for *arrowprops* to customize the arrow color and direction. The final step is some housekeeping to label the axes, create a chart title, create the legend, and show the plot that we have been creating.

```
plt.xlabel('Time (s)')
plt.ylabel('Pressure (psf)')
plt.title('Dynamic pressure as a function of time')
```

```
plt.legend()
plt.show()
```

Hit run, and let's see our creation!

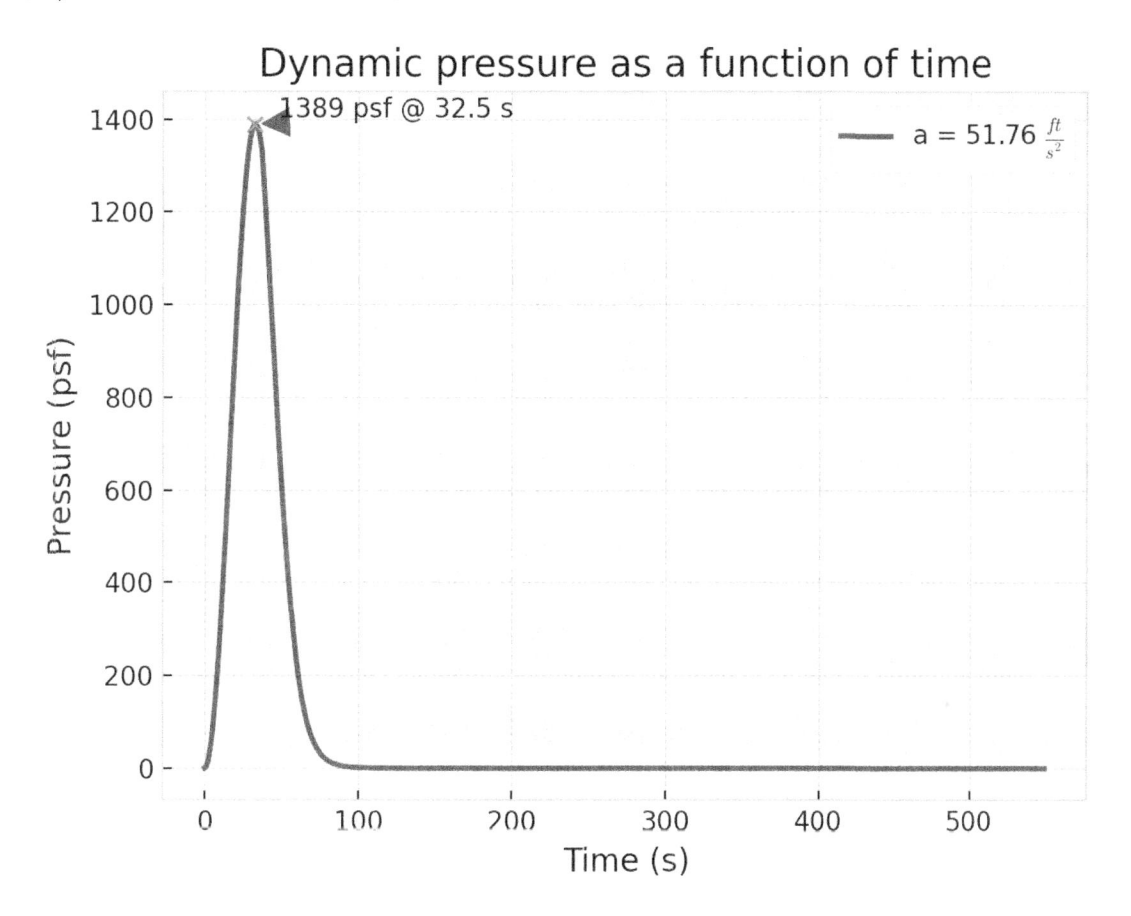

Dynamic pressure as a function of time

1389 psf @ 32.5 s

$a = 51.76 \frac{ft}{s^2}$

Pressure (psf) vs Time (s)

Woah! Notice how Matplotlib automatically formatted both axes to fit all of the data. It looks like we didn't need to go to 550 seconds after all, and it looks like our arrow got squished. We can either re-run the program with something like *np.arange(0.0, 125.0, 0.5)*, or we can limit the x-axis. Let's limit the x-axis. The housekeeping part of code has one new addition:

```
plt.xlim(0, 125) # new!
plt.xlabel('Time (s)')
plt.ylabel('Pressure (psf)')
plt.title('Dynamic pressure as a function of time')
plt.legend()
plt.show()
```

Run the code again and see what you get:

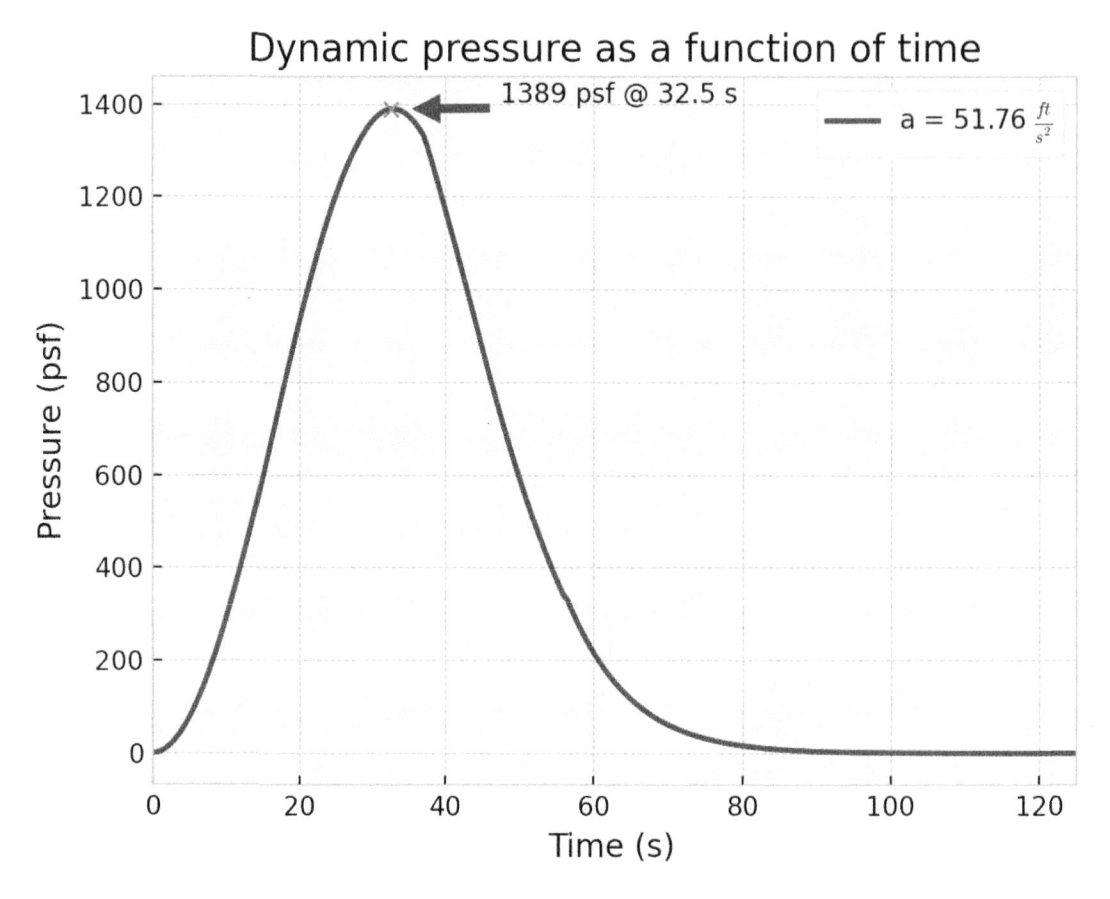

Dynamic pressure as a function of time

1389 psf @ 32.5 s

$a = 51.76 \frac{ft}{s^2}$

Pretty cool! Max Q for a normal rocket launch occurs at about 30 seconds into the launch, and the rocket feels a pressure of 1,400 psf (67 kPa). No wonder the rockets throttle down for Max Q. For reference, sea-level standard pressure is about 2,110 pounds per square foot (101 kPa).

Because of the assumptions we made with constant acceleration, the actual Max Q for a rocket is likely lower. Because of the assumptions we made with constant acceleration, 1 billion psf is most likely nowhere near the Max Q that a rocket experiences; the actual Max Q is likely far lower. On Earth, assuming you are not doing anything crazy like drag racing or flying a fighter jet, you are always experiencing 1 g of acceleration as the Earth rotates. 1 g = 32.2 ft/s^2, so the 51.76 ft/s^2 that we used for average acceleration comes out to about 1.6 g over Earth's natural gravitational pull (2.6 g total). What would a rocket accelerating at 1 g look like on this graph? (In an inertial frame, this is 2 g of acceleration: 1 g from Earth and 1 g going away from Earth.) Since we used acceleration as a variable, all we have to do is copy and paste the code starting with *y_values = [ ]* through *plt.plot()* we used and replace *acceleration = 51.76* with *acceleration = 32.2*. It is good practice to rename *y_values* to *y2_values* or something else that isn't *y_values* so you can preserve both sets of pressures. In Python:

```python
y2_values = []
for elapsed_time in x_values:
    accel = 32.2
    alt = altitude(elapsed_time, accel)
    q = 0.5 * density(alt) * velocity(elapsed_time, accel) ** 2
    y2_values.append(q)

plt.plot(x_values, y2_values, 'k-',
        label=r"a = 32.2 $\frac{ft}{s^2}$")

max_val = max(y2_values)
ind = y2_values.index(max_val)

# Plot an arrow and text with the max value
plt.annotate('{:.0f} psf @ {} s'.format(max_val, x_values[ind]),
```

```
                xy=(x_values[ind] + 3, max_val),
                xytext=(x_values[ind] + 15, max_val + 15),
                arrowprops=dict(facecolor='black', shrink=0.05))

# plot the point of Max Q
plt.plot(x_values[ind], max_val, 'rx')
```

The good news is that we can use the same *plt.plot()* function, and Matplotlib will automatically graph both sets of data. Let's also change the linecolor and linestyle options from *'b-'* to *'k-'* so that we can compare the two datasets. The resulting chart should look like this:

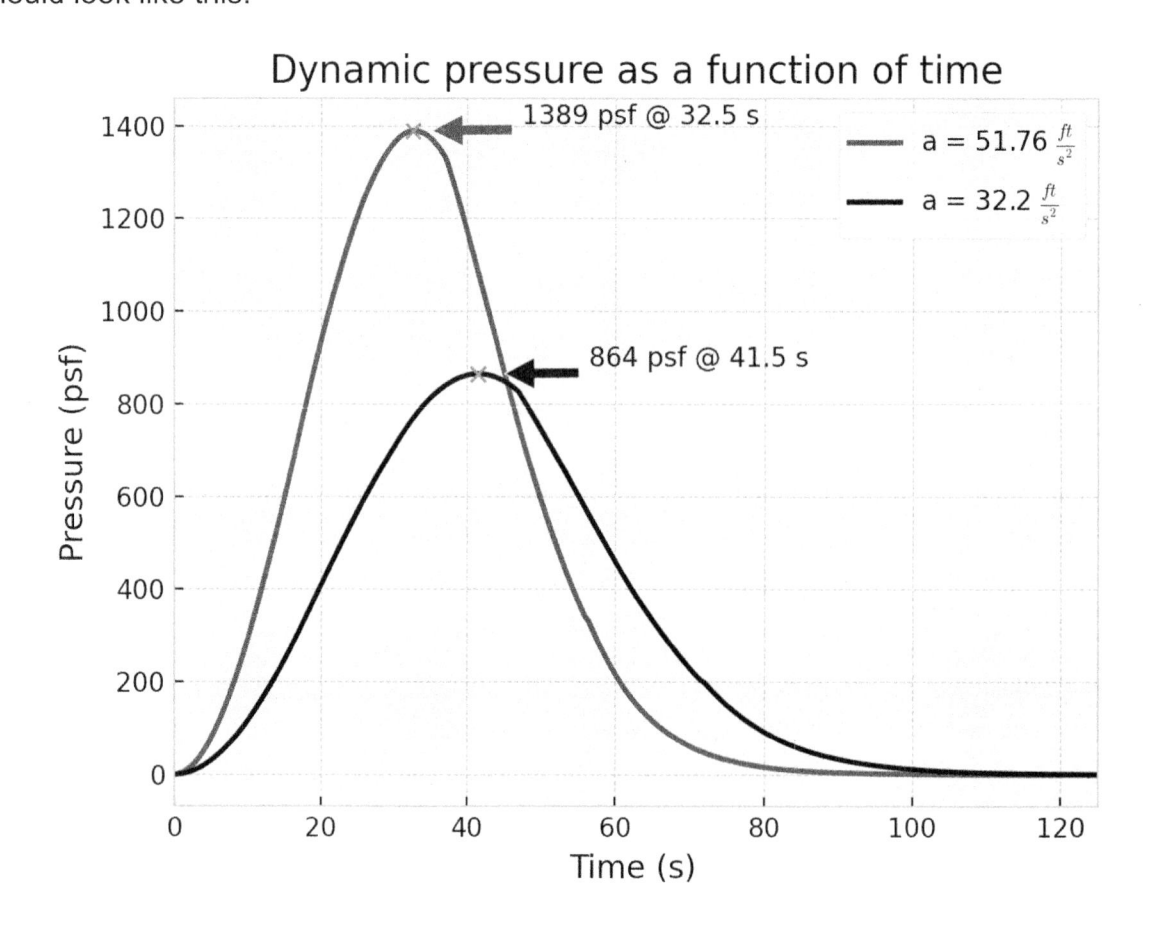

Let's try a third acceleration. A 2016 Wired article estimated that the launch acceleration of the Blue Origin New Shephard rocket is 12.3 ft/s^2 [3]. The rocket will have greater acceleration as fuel is used up and air resistance goes down; let's assume that the rocket's average acceleration from launchpad to orbit is around 20 ft/s^2, which is an estimate that the rocket takes 20 minutes to get to orbital velocity at a constant acceleration. Here is the additional code again:

```
# Now let's make acceleration - 20.0 ft/s^2
y3_values = []
for elapsed_time in x_values:
    accel = 20.0
    alt = altitude(elapsed_time, accel)
    q = 0.5 * density(alt) * velocity(elapsed_time, accel) ** 2
    y3_values.append(q)

plt.plot(x_values, y3_values, 'g-',
         label=r"a = 20.0 $\frac{ft}{s^2}$")
max_val = max(y3_values)
ind = y3_values.index(max_val)

# Plot an arrow and text with the max value
plt.annotate('{:..0f} psf @ {} s'.format(max_val, x_values[ind]),
            xy=(x_values[ind] + 3, max_val),
            xytext=(x_values[ind] + 15, max_val + 15),
            arrowprops=dict(facecolor='green', shrink=0.05))

# plot the point of Max Q
plt.plot(x_values[ind], max_val, 'rx')
```

We will also need to adjust the x-axis limit to 150 so that the legend doesn't overlap the data.

```
plt.xlim(0, 150) # new!
plt.xlabel('Time (s)')
plt.ylabel('Pressure (psf)')
plt.title('Dynamic pressure as a function of time')
plt.legend()
plt.show()
```

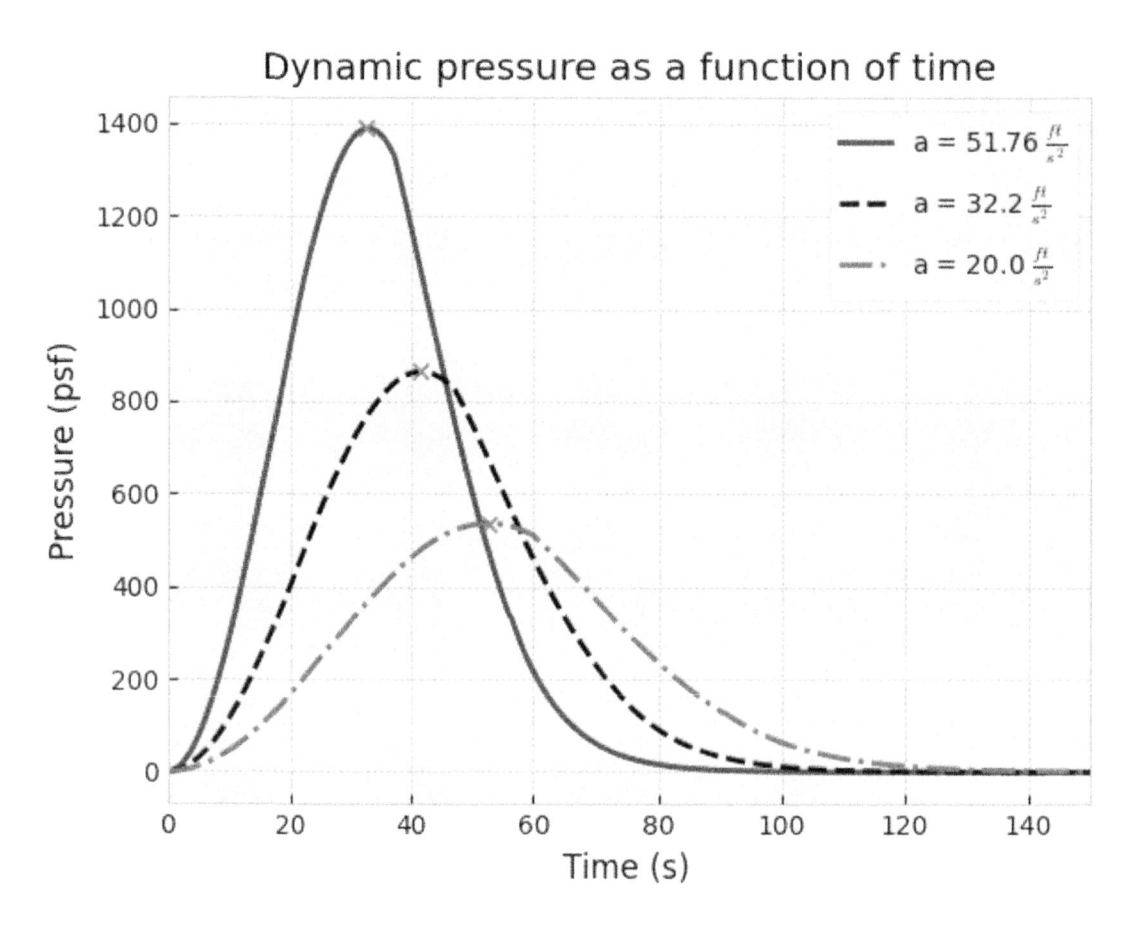

Dynamic pressure as a function of time, with legend entries $a = 51.76 \frac{ft}{s^2}$, $a = 32.2 \frac{ft}{s^2}$, $a = 20.0 \frac{ft}{s^2}$.

The graph shows a drastic relationship between average acceleration and pressure. (The line types have also been adjusted to display better in black and white.) The takeaway here is that changing the launch average acceleration can have dramatic effects on the maximum pressure experienced by the launch vehicle.

Summary

Matplotlib is a great library for building data visualization tools. This example is just a small sample of what the library can help you display and accomplish. Matplotlib also allows you to embed and export the charts to various formats like pdf, svg, and others. Combined with Scipy/Numpy, Matplotlib is a great tool to use to display data in new and interesting ways. We were able to use Matplotlib and some physics to plot maximum aerodynamic pressure, called Max Q, as a function of time. In reality, rockets do not have constant acceleration. The acceleration will vary depending on atmospheric conditions, the altitude of the rocket, the engine performance, and more; however, the assumption of constant acceleration allows us to use simple physics equations to model Max Q. The assumptions are a good way to get a first-order guess onto the graph to show what Max Q is and why it occurs. In later chapters, we will be able to animate graphs with Matplotlib.

This graph and iterative programming process also demonstrates a core tenet of engineering: even if you have all of the mathematical models possible, you still need to sanity-check the output to verify that it makes sense. If you put garbage into a model or equation that you do not understand, you will most likely get garbage out of it, too.

The below Full Code uses a Python convention of if $\_\_name\_\_$ $==$ $"\_\_main\_\_":$. Any Python program you write could be imported by a different program if you so desired. If you were to do this, anything at the zero-indent level would get run on import. $main$ is a hidden method that is invoked whenever you run the program from the text editor area. In essence, this *if* statement is insurance against only running the main part of the program when you mean to. Don't worry too much about it because the program would still run the same if you delete that *if* statement and re-indent everything to the zero-indent level.

Full Code

```python
# Imports
import numpy as np
import matplotlib.pyplot as plt


# Equations
def density(height: float) -> float:
    if height < 36152.0:
        temp = 59 - 0.00356 * height
        p = 2116 * ((temp + 459.7)/518.6)**5.256
    elif 36152 <= height < 82345:
        temp = -70
        p = 473.1*np.exp(1.73 - 0.000048*height)
    else:
        temp = -205.05 + 0.00164 * height
        p = 51.97*((temp + 459.7)/389.98)**-11.388

    rho = p/(1718*(temp+459.7))
    return rho


def velocity(time: float, acceleration: float) -> float:
    return acceleration*time


def altitude(time: float, acceleration: float) -> float:
    return 0.5*acceleration*time**2
```

```python
if __name__ == '__main__':
    plt.style.use('bmh')
    y_values = []
    x_values = np.arange(0.0, 550.0, 0.5)
    for elapsed_time in x_values:
        #
        '''
        Acceleration is the average acceleration
        to go from 0 ft/s to 26,400 ft/s
        (18,000 mph) in 8.5 minutes = 51.764705882 ft/s^2
        '''
        accel = 51.764705882
        alt = altitude(elapsed_time, accel)
        # Dynamic pressure q = 0.5*rho*V^2 = 0.5*density*velocity^2
        q = 0.5 * density(alt) * velocity(elapsed_time, accel) ** 2
        y_values.append(q)

    plt.plot(x_values, y_values, 'b-',
             label=r"a - 51.76 $\frac{ft}{s^2}$")
    max_val = max(y_values)
    ind = y_values.index(max_val)

    # Plot an arrow and text with the max value
    plt.annotate('{:,.0f} psf @ {} s'.format(max_val,
                                              x_values[ind]),
                 xy=(x_values[ind] + 2, max_val),
                 xytext=(x_values[ind] + 15, max_val + 15),
                 arrowprops=dict(facecolor='blue', shrink=0.05),
                 )
    # plot the point of Max Q
    plt.plot(x_values[ind], max_val, 'rx')

    # Now let's make acceleration = 32.2 ft/s^2 (1g)
    y2_values = []
    for elapsed_time in x_values:
        accel = 32.2
        alt = altitude(elapsed_time, accel)
        q = 0.5 * density(alt) * velocity(elapsed_time, accel) ** 2
        y2_values.append(q)
```

```python
plt.plot(x_values, y2_values, 'k-',
         label=r"a = 32.2 $\frac{ft}{s^2}$")
max_val = max(y2_values)
ind = y2_values.index(max_val)

# Plot an arrow and text with the max value
plt.annotate('{:.0f} psf @ {} s'.format(max_val, x_values[ind]),
             xy=(x_values[ind] + 3, max_val),
             xytext=(x_values[ind] + 15, max_val + 15),
             arrowprops=dict(facecolor='black', shrink=0.05))

# plot the point of Max Q
plt.plot(x_values[ind], max_val, 'rx')

# Now let's make acceleration = 20.0 ft/s^2 (1g)
y3_values = []
for elapsed_time in x_values:
    accel = 20.0
    alt = altitude(elapsed_time, accel)
    q = 0.5 * density(alt) * velocity(elapsed_time, accel) ** 2
    y3_values.append(q)

plt.plot(x_values, y3_values, 'g-',
         label=r"a = 20.0 $\frac{ft}{s^2}$")
max_val = max(y3_values)
ind = y3_values.index(max_val)

# Plot an arrow and text with the max value
plt.annotate('{:.0f} psf @ {} s'.format(max_val, x_values[ind]),
             xy=(x_values[ind] + 3, max_val),
             xytext=(x_values[ind] + 15, max_val + 15),
             arrowprops=dict(facecolor='green', shrink=0.05))

# plot the point of Max Q
plt.plot(x_values[ind], max_val, 'rx')

plt.xlim(0, 150)
plt.xlabel('Time (s)')
plt.ylabel('Pressure (psf)')
```

```
plt.title('Dynamic pressure as a function of time')
plt.legend()
plt.show()
```

View this code on GitHub at https://github.com/alexkenan/pymae/

Resources

[1] https://www.nasa.gov/mission_pages/shuttle/shuttlemissions/sts121/launch/qa-leinbach.html

[2] https://www.grc.nasa.gov/www/k-12/rocket/atmos.html

[3] https://www.wired.com/2016/01/how-fast-did-blue-origins-rocket-accelerate-on-launch/

Chapter 4: Getting and Plotting Airfoil Coordinates

Introduction

Airfoils are 2D cross-sections of wings, and they are primarily used to perform first-order modeling of lift and drag at different speeds and angles of attack. Professor Michael Selig at the University of Indiana at Urbana-Champaign (UIUC) has posted an exhaustive list of airfoils and their coordinates at his UIUC website [1]. Airfoils can be designed to maximize or minimize different characteristics, so there are many different airfoil designs.

The Requests library is one of many libraries that can be used to make HTTP and HTTPS website requests. It allows you to interact with websites, submit forms, get text from websites, and more. Though it is not a built-in library, it is one of the most popular libraries to use for web scraping. We will use Requests to help us get airfoil coordinates and plot them with Matplotlib. You can find the Requests documentation link under footnote [2]. Requests is included in the Anaconda distribution, which is one of the reasons that Spyder is recommended.

We will use Requests to make an application programming interface (API) out of the UIUC airfoil website so that we can get any airfoil's coordinates and plot them with Matplotlib. Our program will interface with the UIUC website and give us the coordinates of the requested airfoil. The program should flow as follows:

- Scrape the UIUC website for the specific airfoil coordinates

- Grab the coordinates and split them appropriately so that we have a list of x coordinates and y coordinates
- Do any data cleaning that may be required (removing leading or trailing spaces, getting rid of any tabs, getting rid of any extra or unwanted characters, etc)
- Plot the coordinates

This chapter in particular will develop the code as you read. A lot of programming books go from talking about the program to having the entire program written in one step. There is a lot of troubleshooting and debugging that happens in between the two steps, and being able to debug well is a great skillset to have. See if you notice any problems before they appear, and see if you can think of a better way to write the program as you read.

Imports

```
import requests
import matplotlib.pyplot as plt
```

Requests may not be installed by default, so *pip install requests* as necessary.

Body

First, we will define a *get_airfoil_coordinates()* function that will:
- accept an airfoil name as an argument

- scrape the UIUC website
- get the results and clean the data
- return the x coordinates, the y coordinates, and the airfoil name as elements of a *tuple*

To build this function, we will use the NACA 2412 airfoil to ensure that the program works.

The URL of each airfoil's coordinates in the UIUC airfoil database follows the same format: *https://m-selig.ae.illinois.edu/ads/coord/{airfoil_name}.dat*. Each time we query the database, we will replace *{airfoil_name}* with the airfoil argument in the function. The NACA 2412 file is formatted as follows:

```
NACA 2412
1.0000      0.0013
0.9500      0.0114
0.9000      0.0208
......
0.9500     -0.0048
1.0000     -0.0013
```

Interestingly, it is not a CSV file; rather, it is a raw text (txt) file. The first line of the file is the airfoil name, and the subsequent lines are normalized coordinates in the x and y direction, respectively. The URL also returns a **Not Found** page if any of the airfoil letters are uppercase. We will need to make sure any URL we query uses airfoils with all lowercase letters. Next, in order to figure out what character was separating the two coordinates, we can copy the first line of coordinates, go to the Notebook area, create a new *str* with that line, and try to see what characters would properly split the *str*. The

Tab character is represented as \t in Python; similarly, the new line character (otherwise known as hitting Enter) is represented as \n.

```
>>> text = '1.0000     0.0013'
>>> text.split('\t')
['1.0000     0.0013']
```

Well, that didn't work. There are five spaces separating the numbers, so let's try that:

```
>>> text.split('     ')
['1.0000', '0.0013']
```

Okay, five spaces it is. Now, we can start creating the function. Ultimately, we will have this function return a *tuple*, which is like an unchangeable *list*.

```
def get_airfoil_coords(airfoil: str) -> tuple:
    url = 'https://m-selig.ae.illinois.edu/ads/coord/{}.dat'.format(
                                    airfoil.lower())
    response_text = requests.get(url).text
```

response_text has all of the coordinates as a big *str*, so we'll split it against the new line character "\n" to get a list with each line of the *str* as an element in the *list*. Then, we'll iterate through each element of the *list*, split the text into the x and y coordinates, and add them to master *lists* to remember them for later.

```
    all_text = response_text.split('\n')
    x_coordinates, y_coordinates = [], []
    plot_title = ''

    for line in all_text:
        line = line.strip()
        if len(line.split('     ')) == 1:
```

```
            plot_title = line
        else:
            x, y = line.split('     ')
            x_coordinates.append(x.strip())
            y_coordinates.append(y.strip())
    return x_coordinates, y_coordinates, plot_title
```

We'll figure out the airfoil title to use in the title of the plot by making use of the fact that it will be the only line to have a *list* of length 1. If you try to *split()* a *str* against a character that is not in the *str*, you get a *list* of length 1 with the entire *str* again. Let's see an example:

```
>>> text = 'this is some text'
>>> my_list = text.split('string that is not here')
>>> my_list
['this is some text']
```

(We used this characteristic above to figure out that 5 spaces separated the x and y coordinates.) Therefore, if you try to split a line by five spaces and the resulting *list* is of length 1, then we have the airfoil title. We could also just use the airfoil name; however, there are airfoils that have file names like naca16012, but the airfoil name is stylized like NACA 16-012. If we grab the name from the resulting data, we can just use the stylized name without having to think about it.

Next, we'll create a *plot_airfoil()* function that accepts a *list* of x coordinates, a *list* of y coordinates, the airfoil name as plot title, and displays the airfoil plot.

```
def plot_airfoil(x_coordinates: list, y_coordinates: list,
                 plot_title: str) -> None:
    plt.plot(x_coordinates, y_coordinates, 'k-')
    plt.title('{} airfoil'.format(plot_title))
    plt.grid(None)
    plt.show()
```

Because *plt* is defined at the zero-indentation level in the import section, the *plot_airfoil()* function can find *plt* even if we don't pass it as an argument. However, it is generally good practice to pass *plt* as an argument.

We need to call the two functions, and let's see the results:

```
air_foil = 'naca2412'
x_values, y_values, title = get_airfoil_coords(air_foil)

plot_airfoil(x_values, y_values, title)
```

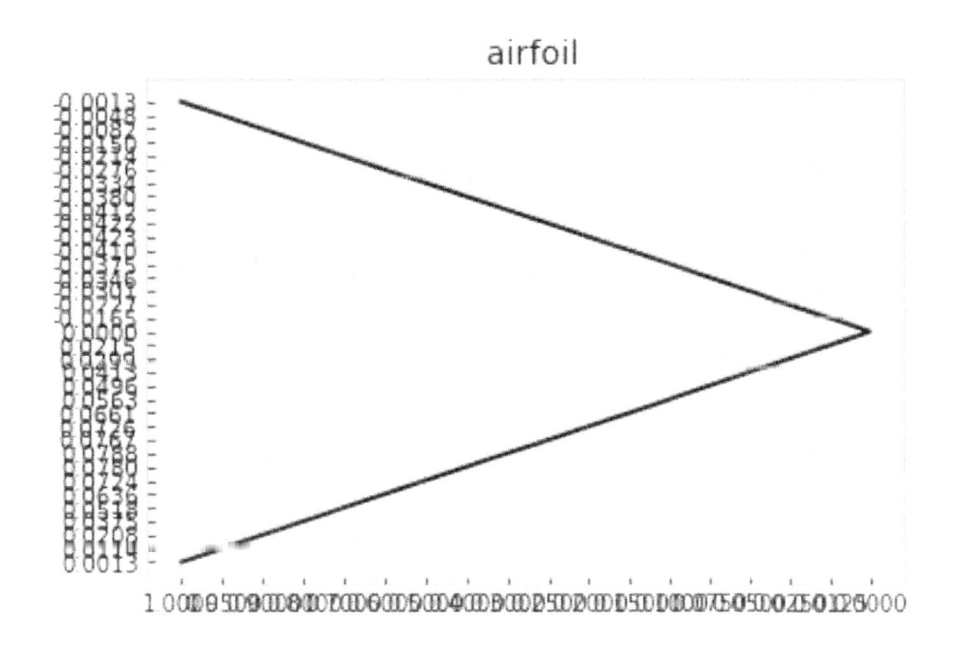

We wanted to graph an airfoil, not a giant greater-than sign. Let's catalog all of the issues:

87

1) The points are not graphed correctly; the points should look like an airfoil

2) The x-axis and y-axis are flipped and the numbers are on top of each other

3) The airfoil name did not display

4) The background should be white

Matplotlib is supposed to automatically graph (left to right) from negative x to positive x, and (down to up) negative y to positive y. Basically, Matplotlib is supposed to automatically graph like a normal graph. Fortunately, addressing issue 1 will also address issue 2. We can address issues 3 and 4 as well.

Since we are pulling the airfoil coordinates from a website and getting the coordinates as *strs*, we might be having an issue where Matplotlib is trying to graph the *str* of the coordinates, not the coordinates as actual numbers. We can solve this issue by casting each x and y value as *float(x.strip())* and *float(y.strip())* instead of just writing *x.strip()* and *y.strip()*. It is also common to run into issues when splitting a text file at the beginning and/or end of the file. In our case, we clearly have at least one issue at the end: we definitely assign *plot_title* to the first line of the file, but it displayed on the graph as an empty *str*. There are at least three ways to solve the problem:

1) Add an additional *if* statement to verify that line contains more than an empty *str* (Remember how an empty *str* is inherently *False* in Python?)

2) Add a *found_title* boolean and add an *elif* statement to check if *found_title* is *True*

3) Use *enumerate()* to assign the plot title if the index is 0 (the first line we read) and use a try-except clause to catch the *ValueError* if the split tries to happen.

All three options are shown below:

```
# Option 1
for line in all_text:
    line = line.strip()
    if line:
        if len(line.split('     ')) == 1:
            plot_title = line
        else:
            x, y = line.split('     ')
            x_coordinates.append(float(x.strip()))
            y_coordinates.append(float(y.strip()))
# Option 2
found_title = False
for line in all_text:
    line = line.strip()
    if len(line.split('     ')) == 1 and not found_title:
        plot_title = line
        found_title = True

    else:
        x, y = line.split('     ')
        x_coordinates.append(float(x.strip()))
        y_coordinates.append(float(y.strip()))
# Option 3
for index, line in enumerate(all_text):
    if index == 0:
        plot_title = line.strip()
    else:
        try:
            line = line.strip()
            x, y = line.split('     ')
            x_coordinates.append(float(x.strip()))
            y_coordinates.append(float(y.strip()))
        except ValueError:
            continue
```

We will use Option 3 going forward because I think it will future-proof the program the best.

Instead of using *plt.grid(None)* in the *plot_airfoil()* function, change it to *plt.style.use('default')*. Now, let's try running the program again:

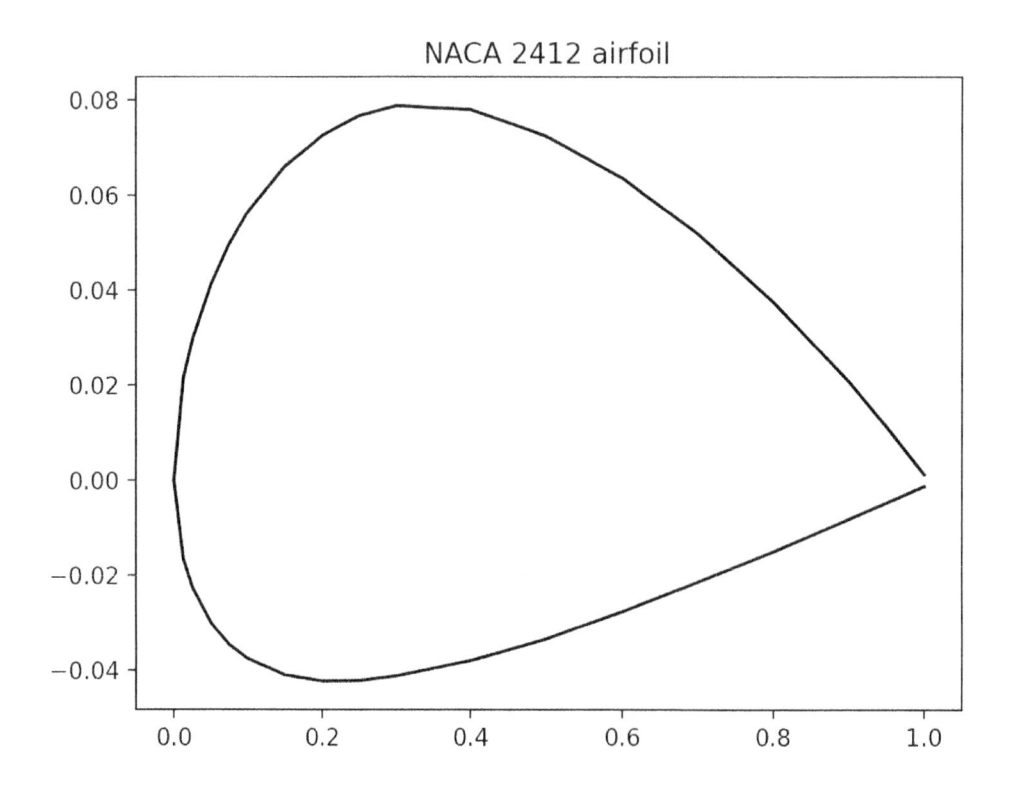

NACA 2412 airfoil

Hey look, it's an airfoil! It's a little bloated, but at least it doesn't look like a greater-than sign anymore. Since the x coordinates are normalized to be between 0 and 1, and the y coordinates are normalized to be between -1 and 1, we can limit the x- and y-axis in the *plot_airfoil()* function with:

```
plt.xlim(-0.50, 1.25)
plt.ylim(-1, 1)
```

Running the script again results in:

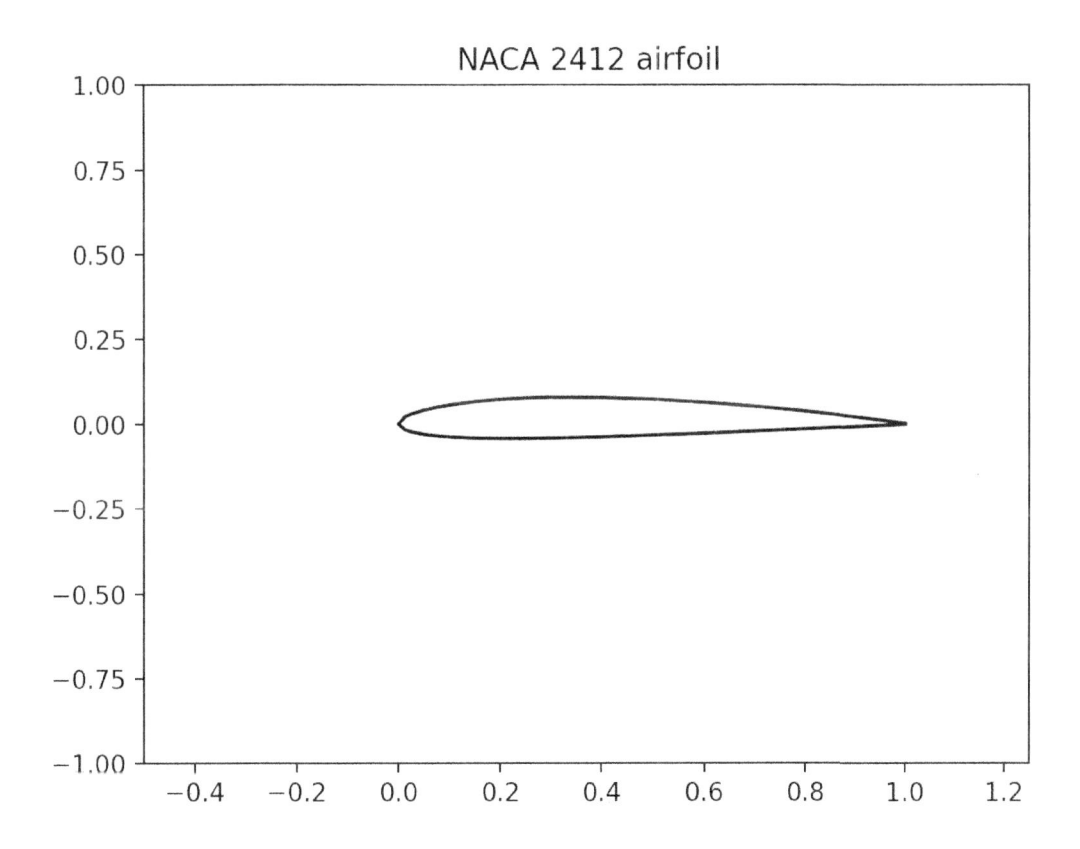

NACA 2412 airfoil

Now, we can graph any NACA airfoil by using its name. But, there are many more airfoils in the UIUC database than just NACA ones. Let's try graphing a MH70 airfoil:

MH 70 11.08% airfoil

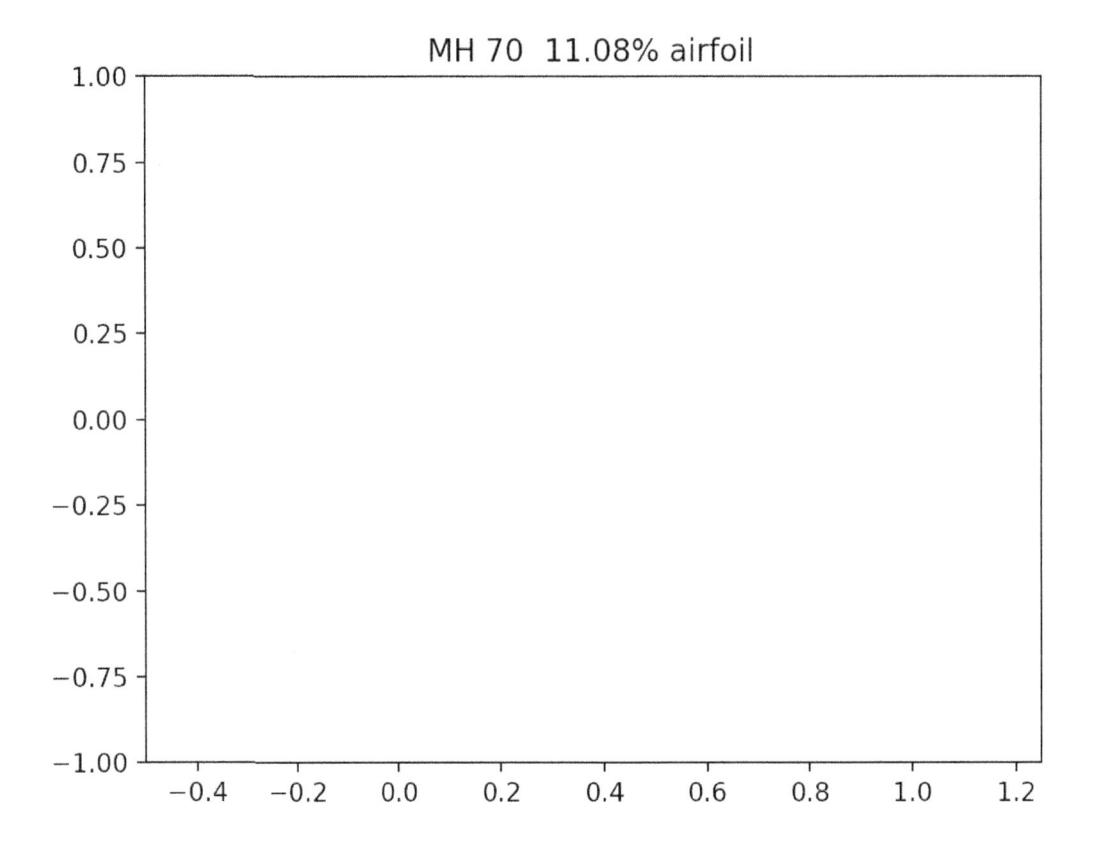

(blank plot with x-axis from −0.4 to 1.2 and y-axis from −1.00 to 1.00)

It's blank. Let's look at the *.dat* file to see why:

```
MH 70   11.08%
  1.00000000  0.00000000
  0.99663874  0.00038973
  0.98681861  0.00160039
. . . . . .
  0.98657712  0.00013332
  0.99660243  0.00006156
  1.00000000  0.00000000
```

The coordinates are separated by two spaces, not five. There are also leading spaces to the x coordinate, but the *strip()* method that we call should take care of that. Some airfoil dat files have numbers greater than 1.0; in addition to casting the *str* as *float*, we should also make sure that the *float* is less than or equal to 1.0. We need to re-write the *for* loop to be able to handle any number of spaces separating the two coordinates. Fortunately, we can use the built-in *count()* method to count the number of spaces in the line, and we can use *str* multiplication to split the line by the number of spaces that *count()* counts.

```
for index, line in enumerate(all_text):
    if index == 0:
        plot_title = line.strip()
    else:
        try:
            line = line.strip()
            x, y = line.split(' ' * line.count(' '))   # line changed
            x = float(x.strip())
            y = float(y.strip())
            if x <= 1.0 and y <= 1.0:                  # line changed
                x_coordinates.append(x)
                y_coordinates.append(y)
        except ValueError:
            continue
```

Because we just want to catch the *ValueError* without doing anything, we use *continue* to tell the Python interpreter to resume the *for* loop. If we add the above code and try the MH70 airfoil again, we should be able to tackle any number of spaces separating the two coordinates:

[Plot: MH 70 11.08% airfoil]

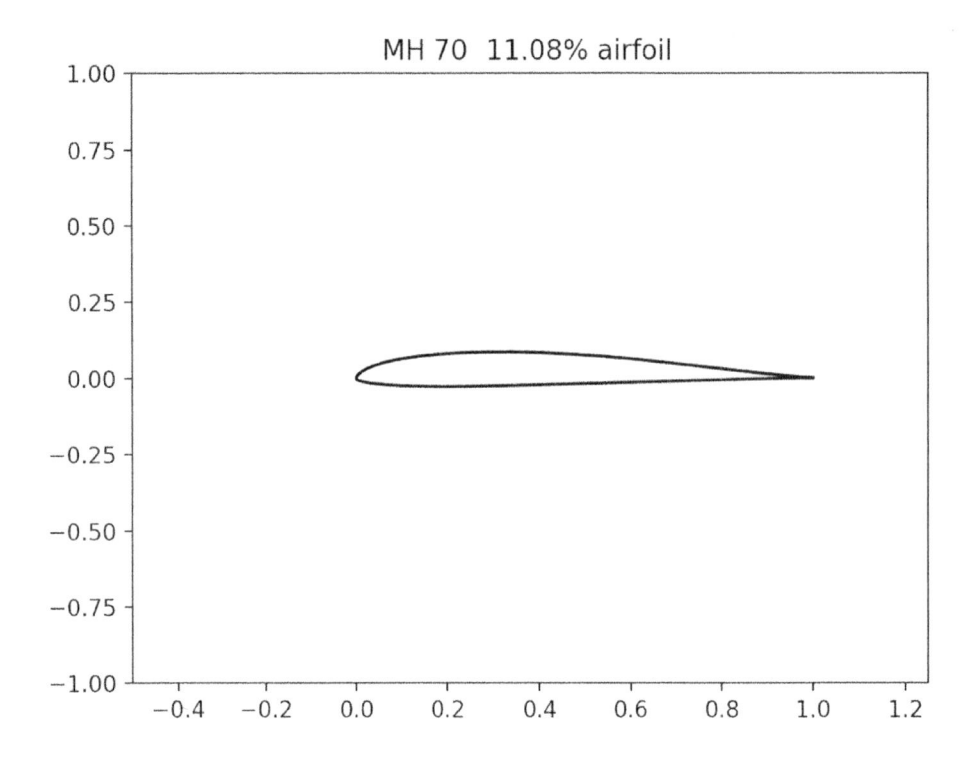

Now our program can handle any airfoil in the UIUC database. But what if we ask it to plot an airfoil that is not in the database?

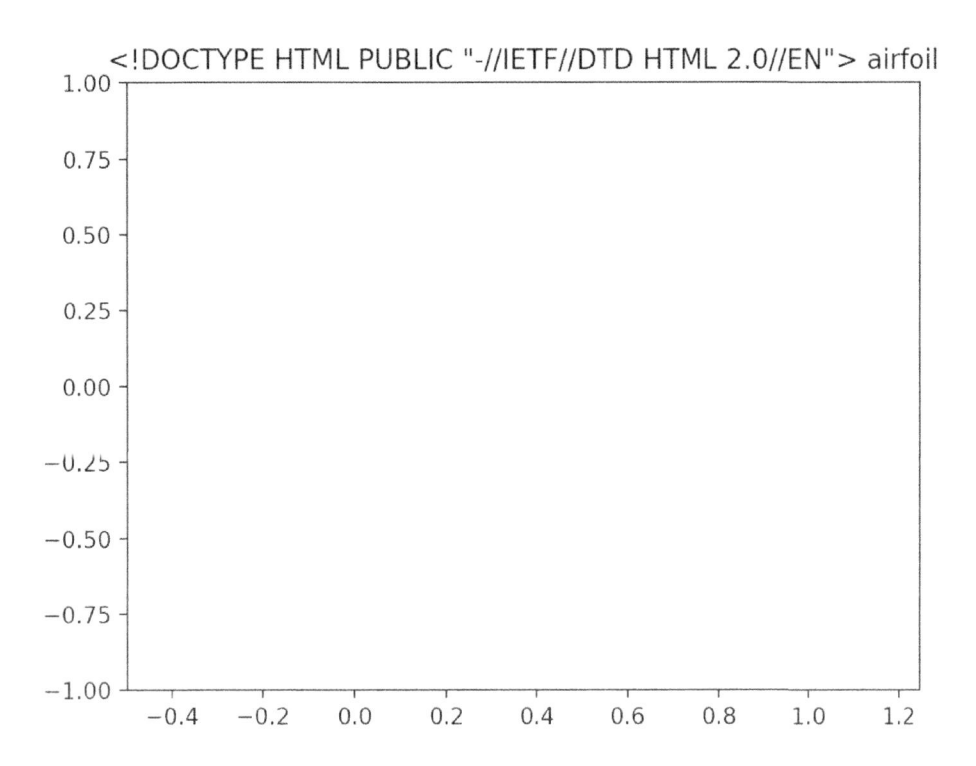

[Plot titled: `<!DOCTYPE HTML PUBLIC "-//IETF//DTD HTML 2.0//EN"> airfoil`, with empty axes ranging from −0.4 to 1.2 on x-axis and −1.00 to 1.00 on y-axis.]

Oops, that's a little embarrassing. If we request a URL that does not exist, the website returns a webpage with big, bold **Not Found** at the top. We can screen for this "not found" piece of text, and if found, we can raise an exception to alert us that the airfoil we requested did not exist. We could create a custom *Exception* or *Error*, but let's just use the built-in *NameError* instead.

In the *get airfoil_coords()* function, we'll add the following if statement after the *response.text* line:

```
if 'Not Found' in response_text:
    raise NameError('{} not found in UIUC database'.format(airfoil))
```

Be careful on the capitalization since the website returns "Not Found" and not "NOT FOUND" or "not found." In the main program, we can couch the script with a try-except clause:

```
air_foil = 'naca2512'
try:
    x_values, y_values, title = get_airfoil_coords(air_foil)
    plot_airfoil(x_values, y_values, title)
except NameError:
    print('{} not in UIUC database, try again!'.format(air_foil))
```

Now we can handle any airfoil in the UIUC database, and we can also handle any fat-fingered typing that the user might input!

Summary

Using the Requests library, we were able to scrape the airfoil coordinates from the UIUC website, clean the data, account for a variable number of spaces separating the two coordinates, and graph the resulting points with Matplotlib. The beauty of Python and its dependent packages is that we can fetch the data with Python (via Requests), clean the data with built-in methods, and display the data (via Matplotlib) all from the same program. If we wanted to, we could even save the coordinates in an Excel spreadsheet with the Openpyxl library and graph the points in Excel. We will get into saving data to Excel in a few chapters.

Full Code

```
import requests
import matplotlib.pyplot as plt


def get_airfoil_coords(airfoil: str) -> tuple:
    url = 'https://m-selig.ae.illinois.edu/ads/coord/{}.dat'.format(
                                                    airfoil.lower())
    response_text = requests.get(url).text

    if 'Not Found' in response_text:
        raise NameError('{} not found in UIUC database'.format(
                                                    airfoil))

    all_text = response_text.split('\n')
    x_coordinates, y_coordinates = [], []
    plot_title = ''

    for index, line in enumerate(all_text):
        if index == 0:
            plot_title = line.strip()
        else:
            try:
                line = line.strip()
                x, y = line.split(' '*line.count(' '))
                x = float(x.strip())
                y = float(y.strip())
                if x <= 1.0 and y <= 1.0:
                    x_coordinates.append(x)
                    y_coordinates.append(y)
            except ValueError:
                continue

    return x_coordinates, y_coordinates, plot_title

def plot_airfoil(x_coordinates: list, y_coordinates: list,
                 plot_title: str) -> None:
    plt.style.use('default')
```

```
        plt.plot(x_coordinates, y_coordinates, 'k-')

        plt.title('{} airfoil'.format(plot_title))
        plt.xlim(-0.50, 1.25)
        plt.ylim(-1, 1)
        plt.show()

air_foil = 'naca2412'
try:
        x_values, y_values, title = get_airfoil_coords(air_foil)
        plot_airfoil(x_values, y_values, title)
except NameError:
        print('{} not in UIUC database, try again!'.format(air_foil))
```

View this code on GitHub at https://github.com/alexkenan/pymae/

Resources

[1] https://m-selig.ae.illinois.edu/ads/coord_database.html

[2] https://requests.readthedocs.io/en/master/

Chapter 5: Modeling a 2-body orbit in 2D and 3D

Introduction

Python is a perfect candidate for modeling simple 2-body orbits like a satellite orbiting the Earth. (It's also a good candidate for complicated 3-body orbits, but we won't get into 3-body orbits in this book.) We want to be able to model basic circular and elliptical orbits in two and three dimensions. We'll take advantage of Matplotlib's *FuncAnimation()* function to animate the 2D orbit to show how the orbital velocity increases at the closest point to Earth and decreases at the farthest point from Earth. We can also use Matplotlib's 3D plotting functionality to plot orbits in 3D, which will help to illustrate the effects of the different orbital parameters.

Solving the 2-body problem requires some form of ordinary differential equations to solve for time and position. These methods usually require a type of numerical method (Newton's Method, Runge-Kutta methods, Kepler's Method, or lots of others) to numerically solve position as a function of time. We could write these differential equation solvers ourselves, but we can use existing libraries that have this functionality. We will use *PyAstronomy*'s *KeplerEllipse* class to create x, y, and z coordinates of a satellite's position over a period of time, then we will use Matplotlib to plot these points. The documentation for PyAstronomy can be found at footnote [1] at the end of this chapter.

The particulars of the hows and whys of orbital mechanics are beyond the scope of this book; however, this paragraph will serve as a crash course in orbital mechanics. A satellite's orbit around a body (we will use Earth in this chapter) can be described and calculated based on the following parameters:

- **Semi-Major and Semi-Minor Axes (a)**: These are the largest and smallest radii measured from the body being orbited to the farthest and closest points in the orbit, respectively. Typically, only the semi-major axis is used as a parameter. For a circular orbit, the semi-major and semi-minor axes are equal, and we call it the radius.

- **Eccentricity (e)**: This is a description of the shape of the orbit. An eccentricity of 0 means the orbit is perfectly circular; an eccentricity between 0 and 1 is an elliptical orbit. For example, an elliptical orbit with eccentricity = 0.50 means that the semi-major axis is twice as large as the semi-minor axis. An eccentricity of 1 is a parabolic orbit, which means that the satellite will slingshot around Earth once and will not return. An eccentricity greater than 1 is a hyperbolic orbit.

- **Inclination (i)**: The angle between the Earth's equator and the angle of the orbit. An orbit with 0 inclination stays over the equator; an orbit with 90 degrees inclination orbits over the North and South Poles. Typically, an orbit's inclination coincides with the latitude at which it was launched: satellites launched from Cape Canaveral, Florida at latitude 30 degrees North have an inclination of 30 degrees.

- **Periapsis and Apoapsis**: Periapsis is the closest point in the satellite's orbit to the body being orbited, and apoapsis is the farthest point in the satellite's orbit to the body being orbited. -apsis is the suffix for a general body, so every orbit has a periapsis. -gee is the suffix for Earth, so all satellites orbiting the Earth have a perigee. -helion is the suffix for the Sun, so all satellites orbiting the Sun have a perihelion.

- **Argument of Periapsis (also called Argument of Perihelion/Perigee) (w)**: The angle between the point on the orbit that intersects the extended equator line and the closest point in the orbit to Earth (perigee). A circular orbit with 0 inclination will have an argument of perigee of 0 degrees. This parameter essentially moves the closest point in the orbit to Earth to a different point along the orbit.

- **Longitude of the Ascending Node (omega)**: The angle between a longitudinal reference (something like the Sun) and the direction of the ascending node, the point when the "climbing" part of the orbit intersects with the extended equator. This parameter can be used to "rotate" the orbit around the Earth.

Don't worry if you don't totally understand all of the parameters. That's what visuals are for. With orbital mechanics out of the way, let's get to Python. We'll start by plotting a 2D static plot of a satellite orbit around Earth, animating the 2D plot, creating a 3D plot of the same orbit, and finally, animating the 3D plot.

Imports

```
import numpy as np
from PyAstronomy import pyasl
import matplotlib.pyplot as plt
import matplotlib.animation as animation
import mpl_toolkits.mplot3d.axes3d as p3
```

PyAstronomy is not part of the default Anaconda distribution, so you will have to download it with *pip install PyAstronomy*.

Body

The beauty of using a third party library (that has been properly vetted) is that the library can handle the hard work, and we only need one line to create the elliptical orbit object *KeplerEllipse*. We'll start with a basic ellipse: semi-major axis of 1.0 units, a time period of 1.0 units, an eccentricity of 0.50, an inclination of 10 degrees, and everything else 0. We'll also use Numpy to create time intervals to be used in determining position later.

```
orbit = pyasl.KeplerEllipse(a=1.0, per=1.0, e=0.5, Omega=0.0,
                            i=10.0, w=0.0)
t = np.linspace(0, 4, 200)
```

The *linspace()* function accepts a starting point, ending point, and a number of samples that you want in between. In this case, it calculates 200 samples between 0 and

4 (inclusive). It means that we don't have to calculate the actual step size; Numpy will do that for us.

All that is left to do is call the *xyzpos()* method, which accepts a time array argument and returns a 3 column array (basically a *List*) the same length as the input time array. It is important to note that the returned array is in the format Y coordinates, X coordinates, Z coordinates. For a 2D plot, we'll ignore the Z coordinates, so our viewpoint is looking down at the North Pole of Earth from above Earth.

```
pos = orbit.xyzPos(t)
```

Congratulations, that's all of the orbital mechanics that is required! The rest of the program will deal with Matplotlib and setting the proper parameters. Because of the convention used within PyAstronomy, the first point in *pos* corresponds to perigee, the closest point to the Earth. We will plot perigee separately so we have our own reference point when we tweak the orbital parameters. The array-like notation that is contained within *pos* means that we will have to use a peculiar double colon :: to get all of the elements within that array for that coordinate system. Earth will be at *(0, 0, 0)* in this coordinate system, and we'll plot it with a big marker size so that we can see it on the plot.

```
plt.plot(0, 0, 'bo', markersize=9, Label="Earth")
plt.plot(pos[::, 1], pos[::, 0], 'k ', Label="Satellite Trajectory")
plt.plot(pos[0, 1], pos[0, 0], 'r*', Label="Periapsis")
```

Since the *pos* object's array coordinates are in the system (Y, X, Z), to get the X coordinates, we need to call the "first" (in reality, the second) element in the array. Add a legend and a title, and let's show the plot:

```
plt.legend(loc="upper right")
plt.title('Orbital Simulation')
plt.show()
```

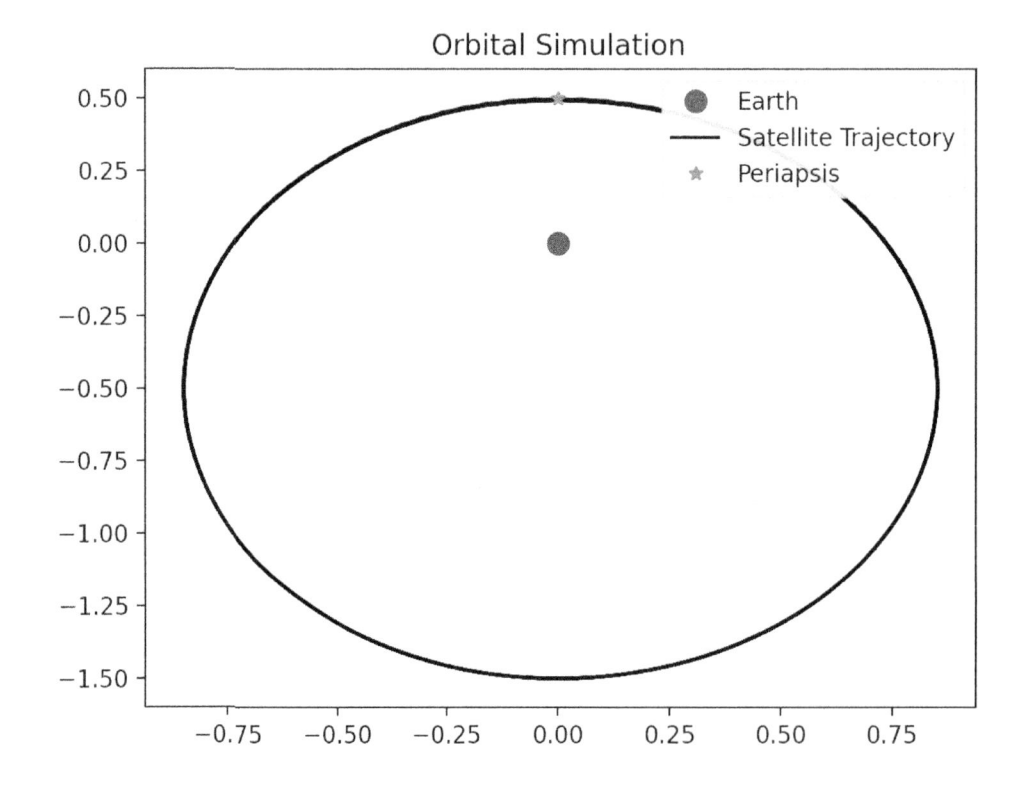

We can see that the orbit looks roughly elliptical, the distance from the Earth to the point farthest away from Earth on the orbit (called apogee) looks to be about 1.5 units, which is consistent with a semi-major axis of 1.0 units and an eccentricity of 0.50. Let's run the program again, but we'll hide all of the tick marks and numbers (since they are normalized to our inputs), and let's give an additional argument of periapsis of 90 degrees to graphically show how that parameter affects the orbit.

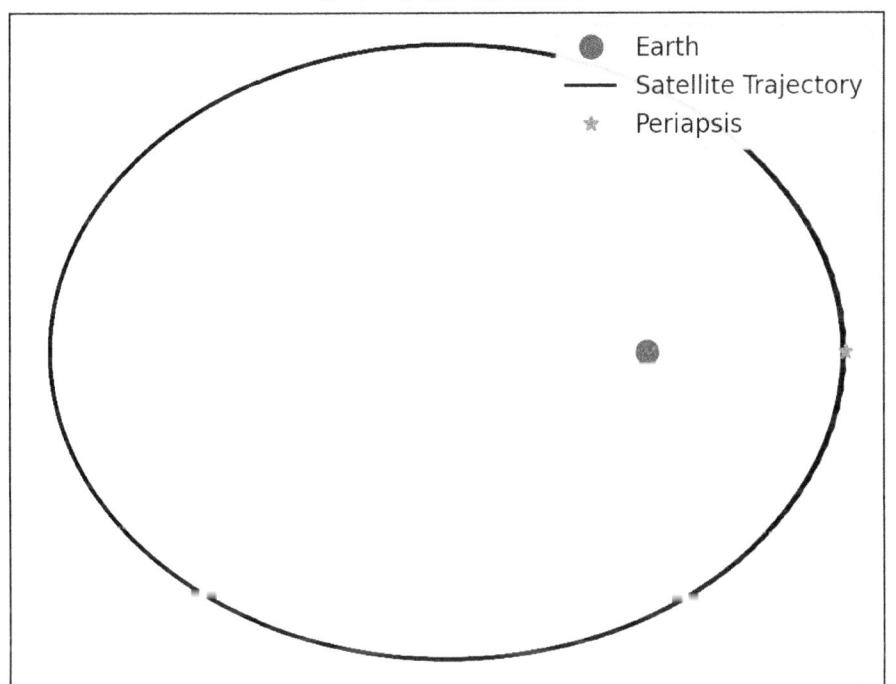

```
orbit = pyasl.KeplerEllipse(a=1.0, per=1.0, e=0.5,
                            Omega=0.0, i=10.0, w=90.0)
t = np.linspace(0, 4, 200)
pos = orbit.xyzPos(t)
plt.legend(loc="upper right")
plt.title('Orbital Simulation')
plt.tick_params(
    axis='both',          # changes apply to both axes
    which='both',         # both major and minor ticks are affected
    bottom=False,         # ticks along the bottom edge are off
    top=False,            # ticks along the top edge are off
    labelbottom=False,    # labels along the bottom edge are off
    left=False,
    labelleft=False)
plt.show()
```

Orbital Simulation

Perigee moved 90 degrees clockwise given a 90 degree argument of perigee.

Now, let's animate the orbit. Note that we have reset the argument of perigee to be 0 degrees. Matplotlib's *FuncAnimation()* takes a function and computes the next value given a range, plots the new value, waits a specified period of time, and then plots the next value. In our case, our animation is really just displaying the pre-computed X and Y coordinates of the orbit. We need to do some prep work to plot the orbit, plot the starting point of a red dot that will go around the orbit, and internally define an *animate()* function that accepts the incremental value *i* and returns *(pos[i][1], pos[i][0])*, so that the *FuncAnimation()* function can plot the new coordinates. Since *pos[i]* will give us the entire *List* of X or Y coordinates, we use the second *[1]* to get the X coordinates or *[0]* to get the Y coordinates of the nested *List*.

```
fig, ax = plt.subplots()

# Plot the orbit (in cyan this time)
plt.plot(pos[::, 1], pos[::, 0], 'c-.')

# Create the red dot as a tuple, and start it at perigee
red_dot, = plt.plot(pos[0][1], pos[0][0], 'ko')
# note the "," after red_dot which creates a tuple

def animate(i):
    red_dot.set_data(pos[i][1], pos[i][0])
    return red_dot,

# create animation using the animate() function
myAnimation = animation.FuncAnimation(fig, animate,
                    frames=np.arange(0, len(t), 1), interval=40,
                    blit=True, repeat=True)
```

The comma is important after *red_dot*; the comma means that *red_dot* is treated as a *tuple* and not just as a *list* of *lists*. *FuncAnimation* takes a lot of arguments:

- *Fig* is the plot so the animation knows where to display the animation
- *Animate* is the function that provides the animation
- *Frames* is the source of data (the incremental *i* mentioned earlier) that provides the animation. In this case, since we are iterating through an array, we want a range, as large as the original time array, that provides *ints* so that the animation can iterate through each element of the array
- *Interval* is the time delay between each iteration of the animation
- *Blitting* is a 2D graphics technique that separates the background of the plot from the moving red dot. If this argument is *True*, the background is frozen in place, and only the red dot is re-animated each time. It should make the animation load and run faster because it is only re-computing the red dot's coordinates, not re-displaying the entire screen
- *Repeat* asks if the animation should play once and freeze, or continue indefinitely. There is also an optional *repeat_delay* parameter that can delay the repeat by the specified number of seconds

There are a few other optional arguments that we did not use [2]. Finally, we will finish up with the plotting cleanup and running the code.

```
plt.plot(0, 0, 'bo', markersize=9, label="Earth")
plt.tick_params(
    axis='both',   # changes apply to both axes
    which='both',  # both major and minor ticks are affected
    bottom=False,  # ticks along the bottom edge are off
    top=False,     # ticks along the top edge are off
    labelbottom=False,  # labels along the bottom edge are off
    left=False,
    labelleft=False)
plt.legend(loc="upper right")
plt.title('Orbital Simulation')
plt.show()
```

Running this in Spyder will just show a static screenshot (like below). Go to Python > Preferences > iPython Console > Graphics > Select *Automatic* instead of *Inline*. You will need to restart the kernel (right click where it says *Console 1/A* near the variable display area and select *Restart kernel*), or just close out of Spyder and re-open it, for the changes to take effect. After doing this, the plot should pop up in a new window or application.

Orbital Simulation

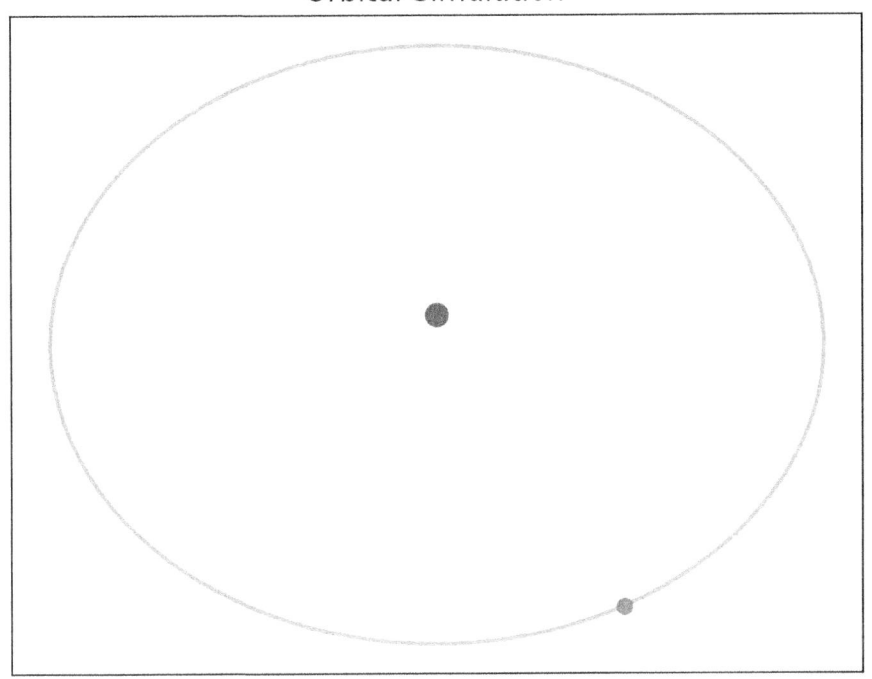

This is only a static screenshot, but the red dot should be moving around the orbit. Congratulations on your first animated plot! If we wanted to be really funny, we could use Matplotlib's *dark_background* style and re-color the dot so that the background is black, the dot is white, and it almost looks like space. Using the dark background brings out the previously-hidden white gridlines, so we will need to turn those off, too!

```
plt.style.use('dark_background')
fig, ax = plt.subplots()
l = plt.plot(pos[::,1], pos[::,0], 'k-')

red_dot, = plt.plot(pos[0][1], pos[0][0], 'wo')

def animate(i):
    red_dot.set_data(pos[i][1], pos[i][0])
```

```python
    return red_dot,

# create animation using the animate() function
myAnimation = animation.FuncAnimation(fig, animate,
                    frames=np.arange(0, len(t), 1), interval=40,
                    blit=True, repeat=True)

plt.plot(0, 0, 'bo', markersize=9, label="Earth")
plt.tick_params(
    axis='both',
    which='both',
    bottom=False,
    top=False,
    labelbottom=False,
    left=False,
    labelleft=False)
plt.axis('off')
plt.legend(loc="upper right")
plt.title('Orbital Simulation')
plt.show()
```

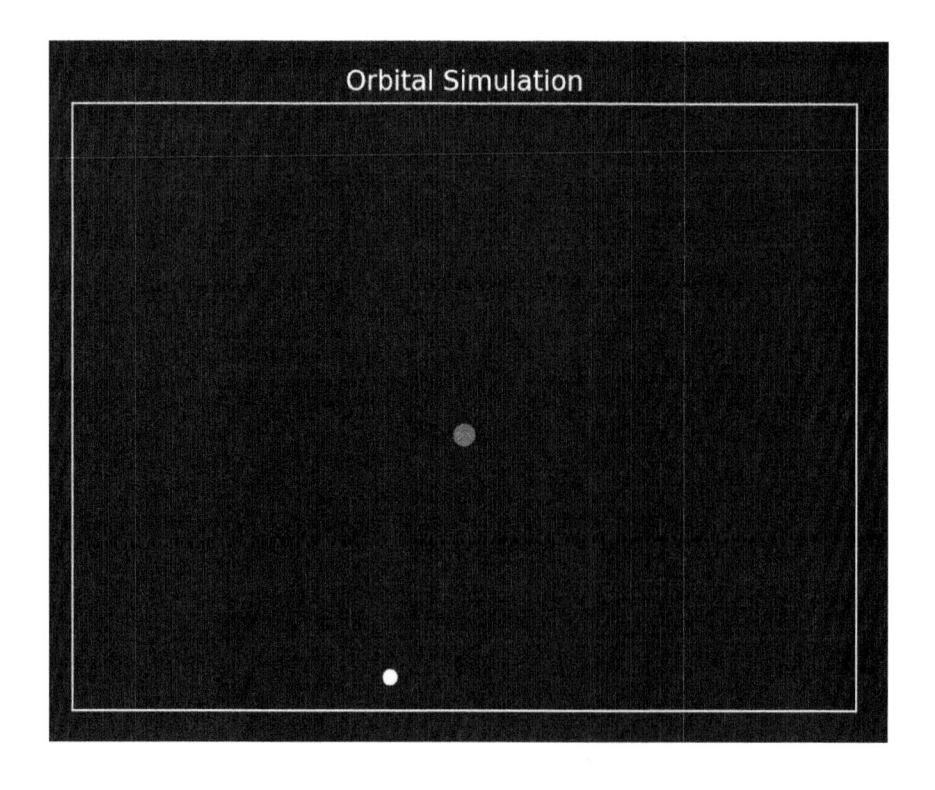

Now, we need to tackle the 3D plot. Graphing in 3 dimensions is a little more complicated, but we already have the Z coordinates of the orbits that we have previously been ignoring. We will need to call a 3D axis projection and plot with the axis object, not with *plt*. We also need to take care to supply coordinates as something that is array-like. This will not be a problem for graphing the coordinates of the orbit, but it might be a problem when trying to plot a specific point like the Earth *(0, 0, 0)* or perigee. Also, be sure to supply the third coordinate when calling the *plot()* method. We want three dimensions, after all. We need to change the code to get rid of the tick marks on the 3D planes, and we need to reset the style to default.

```
orbit = pyasl.KeplerEllipse(a=1.0, per=1.0, e=0.50, Omega=0.0,
                            i=30.0, w=0.0)
# Get a time axis
t = np.linspace(0, 4, 200)

# Calculate the orbit position at the given points
# in a Cartesian coordinate system.
pos = orbit.xyzPos(t)

fig = plt.figure()
ax = plt.axes(projection='3d')

ax.plot3D([0], [0], [0], 'bo', markersize=9, label="Earth")

# Plot the satellite orbit
ax.plot3D(pos[::, 1], pos[::, 0], pos[::, 2], 'k-',
          label="Satellite Trajectory")

# Point of periapsis
ax.plot3D([pos[0, 1]], [pos[0, 0]], [pos[0, 2]], 'r*',
          label="Periapsis")

plt.legend(loc="lower right")
```

```
# Hide grid lines
ax.grid(False)

# Hide axes ticks
ax.set_xticks([])
ax.set_yticks([])
ax.set_zticks([])
plt.style.use('default')
plt.title('Orbital Simulation')
plt.show()
```

Orbital Simulation

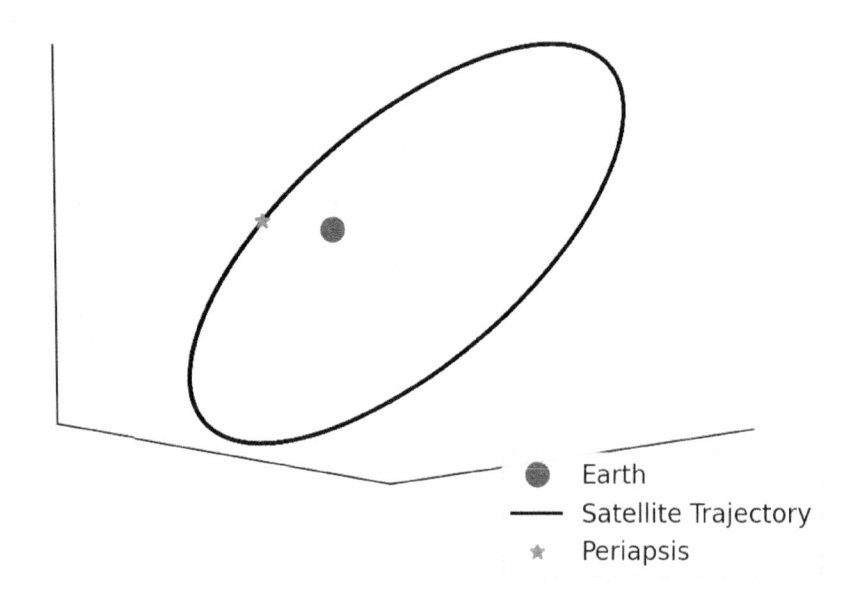

Matplotlib allows you to pan around the plot and to zoom in and out; there are also different styles available [3] for use besides the *default* and *dark_background* styles shown above. You could even make a 3D space plot similar to the 2D space plot created above. This graphic highlights how inclination affects orbits, which you wouldn't be able to see in a 2D plot.

112

And now, the grand finale (for this chapter): the 3D, animated plot. It is a mix of the 2D and 3D code:

```
fig = plt.figure()
ax = p3.Axes3D(fig)

red_dot, = ax.plot(pos[::, 1], pos[::, 0], pos[::, 2], 'ro',
                   label="Satellite")
def animate(i, pos, red_dot):
    red_dot.set_data([pos[i][1], pos[i][0]])
    red_dot.set_3d_properties(pos[i][2])
    return red_dot,

# create animation using the animate() function
ani = animation.FuncAnimation(fig, animate, 200,
                              fargs=(pos, red_dot),
                              interval=100, blit=False)

ax.plot([0], [0], [0], 'bo', markersize=9, label="Earth")

ax.plot(pos[::, 1], pos[::, 0], pos[::, 2], 'k-',
        label="Satellite Trajectory")

# Hide grid lines
ax.grid(False)

# Hide axes ticks
ax.set_xticks([])
ax.set_yticks([])
ax.set_zticks([])
plt.style.use('default')
plt.legend()
plt.show()
```

A static screenshot will not do the code justice, but you should now have a 3D, animated plot moving on your screen. But how do you save it to show off to someone later?

113

For any of the static plots, the Matplotlib window shows a save button to save the image. If you want to save a gif of the animated plots, you need to create a *Writer* instance using Pillow, then pass that *Writer* to a *save()* function:

```
from matplotlib.animation import PillowWriter
writer = PillowWriter(fps=30) # how many frames per second
myAnimation.save('/Users/Alex/Desktop/myAnimation.gif',
            writer=writer)
```

Here is the full code for a saving a gif of the 2D "space" orbit from above:

```
import numpy as np
from PyAstronomy import pyasl
import matplotlib.pyplot as plt
import matplotlib.animation as animation
from matplotlib.animation import PillowWriter
writer = PillowWriter(fps=30)

orbit = pyasl.KeplerEllipse(a=1.0, per=1.0, e=0.50, Omega=0.0,
                  i=30.0, w=0.0)
t = np.linspace(0, 4, 200)
pos = orbit.xyzPos(t)
plt.style.use('dark_background')
fig, ax = plt.subplots()
l = plt.plot(pos[::,1], pos[::,0], 'k-')

red_dot, = plt.plot(pos[0][1], pos[0][0], 'wo')

def animate(i):
    red_dot.set_data(pos[i][1], pos[i][0])
    return red_dot,

# create animation using the animate() function
myAnimation = animation.FuncAnimation(fig, animate,
              frames=np.arange(0, len(t), 1), interval=40,
              blit=True, repeat=True)
```

```
plt.plot(0, 0, 'bo', markersize=9, label="Earth")
plt.tick_params(axis='both', which='both', bottom=False, top=False,
                labelbottom=False, left=False, labelleft=False)
plt.legend(loc="upper right")
plt.title('Orbital Simulation')
myAnimation.save('/Users/Alex/Desktop/myAnimation.gif',
                writer=writer)
```

An important piece of code to point out here is how we are passing color arguments. Instead of the *'k-'* that we have used before, we can pass a hex color to get more custom color options with the arguments *color='#FFFFFF', marker='k'*. These hex colors are used to color webpages, and the hex color codes can be found all over the internet. There are a few hex color resources at the end of this chapter [4].

As you adjust some of the orbital parameters, there may be some modeling issues with PyAstronomy. As eccentricities approach 1.0, there may be some wobbles in the model near perigee as perigee converges on Earth. This is partially an issue with PyAstronomy due to the satellite's velocity at a perigee very close to Earth. The model has to discretize over a larger distance (since velocity is higher for the same time step), so you will see a more jagged line. This error can be improved by increasing the number of steps passed to *linspace()* (for example, use *np.linspace(0, 4, 1000)* instead of *np.linspace(0, 4, 200)*). This will have the effect of slowing down the orbit, but it will reduce the jagged orbit lines.

Summary

This chapter provided a brief introduction to orbital mechanics and used an existing third-party library to model 2D and 3D elliptical orbits with Matplotlib. Third party libraries are a great way to let others' expertise guide you along your programming journey. The *KeplerEllipse* class within PyAstronomy clocks in at about 660 lines of code; we were able to import it and plot an ellipse in about 10 meaningful lines of code. This program is a good learning tool to understand the different orbital parameters and graphically display them. Argument of periapsis and longitude of the ascending node are hard to conceptualize with words, and even a static figure does not clearly convey the picture, either. Being able to graphically represent these parameters and watch the orbits change as the parameters change is an invaluable learning tool, and it can be applied to anything from learning the basics to being used in advanced orbital analysis.

With the steps you learned in this chapter, you can now plot any satellite orbit. The Russian *Molniya* orbit is often discussed in orbital mechanics; the Russians wanted a satellite to be overhead for half of the day for communications and remote sensing at high latitudes. This meant that the satellite needed to have a very low perigee and a very high apogee. Plot the *Molniya* orbit based on the following orbital parameters from Wikipedia [5]:

- Semi-major axis (*a*): 26,600 km
- Period: 0.5 days

- Eccentricity (*e*): 0.737
- Longitude of the ascending node (*omega*): 0 degrees
- Inclination (*i*): 63.4 degrees
- Argument of perigee (*w*): 270 degrees

I used MATLAB to create similar plots in an Astrodynamics course. Making the same plot required messing with an ode45 differential equation solver script, and the whole thing was generally a pain. The beauty of Python is its built-in sharing infrastructure and ability to use packages to do the hard parts for you.

Full Code

```python
import numpy as n
from PyAstronomy import pyasl
import matplotlib.pyplot as plt
import matplotlib.animation as animation
import mpl_toolkits.mplot3d.axes3d as p3
orbit = pyasl.KeplerEllipse(a=1.0, per=1.0, e=0.50, Omega=0.0,
                            i=30.0, w=0.0)

# Get a time axis
t = np.linspace(0, 4, 200)

# Calculate the orbit position at the given points
pos = orbit.xyzPos(t)

# Plot x and y coordinates of the orbit with a few different options

# 2D, animated, and space
'''
plt.style.use('dark_background')
fig, ax = plt.subplots()
l = plt.plot(pos[::,1], pos[::,0], 'k-')
```

```python
red_dot, = plt.plot(pos[0][1], pos[0][0], 'wo')

def animate(i):
    red_dot.set_data(pos[i][1], pos[i][0])
    return red_dot,

# create animation using the animate() function
myAnimation = animation.FuncAnimation(fig, animate,
                        frames=np.arange(0, len(t), 1),
                        interval=40, blit=True, repeat=True)

plt.plot(0, 0, 'bo', markersize=9, label="Earth")
plt.tick_params(
    axis='both',       # changes apply to the x-axis
    which='both',      # both major and minor ticks are affected
    bottom=False,      # ticks along the bottom edge are off
    top=False,         # ticks along the top edge are off
    labelbottom=False, # labels along the bottom edge are off
    left=False,
    labelleft=False)
plt.axis('off')
plt.title('Orbital Simulation')
plt.show()
'''

# 2D and space
'''
plt.style.use('dark_background')

# Plot the satellite orbit
plt.plot(pos[::, 1], pos[::, 0], 'w-', label="Satellite Trajectory")

# Point of periapsis
plt.plot(pos[0, 1], pos[0, 0], 'r*', label="Periapsis")

plt.plot(0, 0, 'bo', markersize=9, label="Earth")
plt.tick_params(
    axis='both',       # changes apply to the x-axis
    which='both',      # both major and minor ticks are affected
    bottom=False,      # ticks along the bottom edge are off
```

```python
        top=False,       # ticks along the top edge are off
        labelbottom=False,  # labels along the bottom edge are off
        left=False,
        labelleft=False)
plt.axis('off')

plt.title('Orbital Simulation')
plt.show()
'''

# 2D, static, white background
#'''
# Plot the Earth
plt.plot(0, 0, 'bo', markersize=9, label="Earth")

# Plot the satellite orbit
plt.plot(pos[::, 1], pos[::, 0], 'k-', label="Satellite Trajectory")

# Point of periapsis
plt.plot(pos[0, 1], pos[0, 0], 'r*', label="Periapsis")

plt.style.use('default')
plt.legend(loc="upper right")
plt.tick_params(
    axis='both',     # changes apply to the x-axis
    which='both',    # both major and minor ticks are affected
    bottom=False,    # ticks along the bottom edge are off
    top=False,       # ticks along the top edge are off
    labelbottom=False,  # labels along the bottom edge are off
    left=False,
    labelleft=False)
plt.title('Orbital Simulation')
plt.axis('off')
plt.show()
#'''

# 2D, animated, white background
'''
fig, ax = plt.subplots()
plt.plot(pos[::, 1], pos[::, 0], 'c-')
```

```python
red_dot, = plt.plot(pos[0][1], pos[0][0], 'ro')


def animate(i):
    red_dot.set_data(pos[i][1], pos[i][0])
    return red_dot,

# create animation using the animate() function
myAnimation = animation.FuncAnimation(fig, animate,
                        frames=np.arange(0, len(t), 1),
                        interval=40, blit=True, repeat=True)

plt.plot(0, 0, 'bo', markersize=9, label="Earth")
plt.tick_params(
    axis='both',    # changes apply to the x-axis
    which='both',   # both major and minor ticks are affected
    bottom=False,   # ticks along the bottom edge are off
    top=False,      # ticks along the top edge are off
    labelbottom=False,  # labels along the bottom edge are off
    left=False,
    labelleft=False)
plt.style.use('default')
plt.title('Orbital Simulation')
plt.axis('off')
plt.show()
'''

# 3D, dark background
'''
plt.style.use('dark_background')
fig = plt.figure()
ax = plt.axes(projection='3d')

ax.plot3D([0], [0], [0], 'bo', markersize=9, label="Earth")

# Plot the satellite orbit
ax.plot3D(pos[::, 1], pos[::, 0], pos[::, 2], 'w-',
          label="Satellite Trajectory")
```

```python
# Point of periapsis
#ax.plot3D([pos[0, 1]], [pos[0, 0]], [pos[0, 2]], 'r*',
            label="Periapsis")
plt.legend(loc="lower right")

# Hide grid lines
ax.grid(False)

# Hide axes ticks
ax.set_xticks([])
ax.set_yticks([])
ax.set_zticks([])
plt.title('Orbital Simulation')
plt.axis('off')
plt.show()
'''

# 3D, white background
'''
fig = plt.figure()
ax = plt.axes(projection='3d')

ax.plot3D([0], [0], [0], 'bo', markersize=9, label="Earth")

# Plot the satellite orbit
ax.plot3D(pos[::, 1], pos[::, 0], pos[::, 2], 'k-',
            label="Satellite Trajectory")

# Point of periapsis
ax.plot3D([pos[0, 1]], [pos[0, 0]], [pos[0, 2]], 'r*',
            label="Periapsis")

plt.legend(loc="lower right")
plt.style.use('default')

# Hide grid lines
ax.grid(False)

# Hide axes ticks
ax.set_xticks([])
```

```python
ax.set_yticks([])
ax.set_zticks([])
plt.title('Orbital Simulation')
plt.show()
'''
# 3D, white background, animated
#'''
fig = plt.figure()
ax = p3.Axes3D(fig)

red_dot, = ax.plot(pos[::, 1], pos[::, 0], pos[::, 2], 'ro',
                   label="Satellite")

def animate(i, pos, red_dot):
    red_dot.set_data([pos[i][1], pos[i][0]])
    red_dot.set_3d_properties(pos[i][2])
    return red_dot,

# create animation using the animate() function
ani = animation.FuncAnimation(fig, animate, 200,
                              fargs=(pos, red_dot),
                              interval=100, blit=False)

ax.plot([0], [0], [0], 'bo', markersize=9, label="Earth")
ax.plot(pos[::, 1], pos[::, 0], pos[::, 2], 'k-',
        label="Satellite Trajectory")
# Hide grid lines
ax.grid(False)

# Hide axes ticks
ax.set_xticks([])
ax.set_yticks([])
ax.set_zticks([])
plt.style.use('default')
plt.legend()
plt.show()
#'''
```

View this code on GitHub at https://github.com/alexkenan/pymae/

Resources

[1] https://www.hs.uni-hamburg.de/DE/Ins/Per/Czesla/PyA/PyA/index.html

[2] https://matplotlib.org/api/_as_gen/matplotlib.animation.FuncAnimation.html

[3] https://matplotlib.org/3.1.1/gallery/style_sheets/style_sheets_reference.html

[4] Here are a few options for hex color codes:

- https://htmlcolorcodes.com/
- https://www.w3schools.com/colors/colors_picker.asp
- https://www.color-hex.com/

[5] https://en.wikipedia.org/wiki/Molniya_orbit

Chapter 6: Unit Conversions

Introduction

NASA launched the Mars Climate Orbiter in 1998 to study changes in the Martian atmosphere and climate. In September 1999, NASA lost communication with the orbiter as it was performing an orbital insertion above Mars. It was later discovered that a piece of software provided results in non-SI units, when the specifications sheet and software receiving the calculations expected SI units. The orbiter crashed into Mars as a result of the unit conversion error, and the experiment was prematurely concluded. That is an expensive unit error.

So far, all of our programs have been run within the script editor and any results have been printed to the console. A Graphical User Interface, called a GUI and pronounced gooey, is the more advanced way to have a user interact with your program and see its results. There are several different libraries for creating a GUI in Python; in no particular order, there are PyQT, Tkinter, Wx, Kivy, and many others. Tkinter is a built-in library for Python, so we will use it to create a GUI to convert values in different pressure units. Its documentation [1] and a guide [2] are linked in the Resources section.

Pressure is defined as force per unit area, so the base unit in SI is Newton per meter squared or Pascal. The US unit is pound-force per square inch or psi. We will use Pint, a Python units library, to be able to assign and convert different units. The documentation for Pint can be found under footnote [3]. The GitHub repository for Pint [4] shows all of the default pressure units that we will use later.

This program will go through three iterations: the first time, we will see how building a GUI works, so we will hard-code in the conversion from megapascals, MPa, to pounds per square inch, psi. After that, we will learn how to make a dropdown menu, so we can offer the user the option to convert between any base pressure unit. Finally, we will make a free-form text entry; by doing so, we can harness the full functionality of Pint and let the user convert between any units of the same dimension, like velocity or distance, not just pressure.

Imports

```
import tkinter as tk
import pint
```

Pint is not a default library, so install it with *pip install pint*. Some use the convention *from tkinter import ** which imports everything from Tkinter without having to use the syntax *tkinter.blah*. I prefer to import libraries the usual way (*import tkinter as tk*) to avoid any potential namespace conflicts. This alternative syntax could create weird conflicts if Tkinter had a function called *blah()*, for example, and you unknowingly decided to create a different function called *blah()*.

Body

When making a GUI with Tkinter, you need to decide if you are going to arrange elements by gridding them with *grid()* or packing them with *pack()*. *Pack()* packs the elements sequentially as you create them, while *grid()* allows you to grid according to row number and column number. We will use grid in all of our examples. If you mix grid and pack, your program will either error out, or it will run in an infinite loop but not display anything as it tries to figure out how to display elements that are gridded and packed (hint: it's not possible, so Tkinter will never figure it out).

All of the elements that you can create with Tkinter are still Python objects; they can be manipulated, named, renamed, etc. "Widget" is a popular term that is used to label any object. Below is a table of the common Tkinter widgets:

Tkinter Name	**Description**
Tk.Label()	A stage to display text
Tk.Entry()	A text box where the user can write text
Tk.Button()	A button to click
Tk.RadioButton()	A selection radio button. Choose one option only
Tk.CheckButton()	A selection checkmark button. Choose as many options as you want
Tk.OptionMenu()	A dropdown menu. Choose one option only

There are many more options, but these are usually the beginner's most popular widgets.

The first step is to create the *Tk* instance, set the width and height of the window, and name the window.

```
app = tk.Tk()
app.title('Unit Converter')
app.geometry('350x200')
# dimensions are X by Y which means width and height in pixels
```

Now, we can start making widgets. There are two ways to do this: the object-oriented way we have been doing by assigning variable names, or a dynamic way where we do not assign a variable name.

```
# Example 1
mpa_text = tk.Label(app, text=" MPa")
mpa_text.grid(row=1, column=3)
# Example 2
mpa_text = tk.Label(app, text=" MPa").grid(row=1, column=3)
# Example 3
tk.Label(app, text=" MPa").grid(row=1, column=3)
```

If you dynamically create the widgets like in example 3, once it is created, it can't be edited. Only use this method if you will not be tweaking that widget later in the code. Example 2 should almost never be used because *mpa_text* will be of type *None*, not the *Label* instance.

Using the gridder allows us to lay out the GUI in a table and easily build it with row and column numbers. Here is a mockup of what the GUI should look like:

	Column 1	Column 2	Column 3
Row 1		*Label* saying "Converter"	
Row 2	*Label* saying "Convert"	*Entry* field for inputting a *str*	*Label* saying " MPa"
Row 3		*Label* saying "into"	
Row 4		*Label* that will eventually have the computed result	*Label* saying " psi"
Row 5		*Button* saying "Calculate"	

It makes no difference if you start with row=0 or row=1; Tkinter will automatically assign the lowest number as the start and count up accordingly. This chapter's code will usually start with 1 instead of 0, but this is probably bad practice because of zero-indexing as discussed in the first chapter.

We also need to create a variable using *tk.StringVar()*. This is a *str* variable that we will use to store the user input. When we click the "Calculate" button, the button will pass that *str* variable (and the *tk.Entry()* field that will eventually have the result) to a *calculate()* function that will perform the calculation. We will also use a default font of Times New Roman, size 14, for all widgets. The title in Row=1 and Column=2 will be Times New Roman, size 24.

```
text_font = ('Times New Roman', 14)

tk.Label(app, text="Converter", font=("Times New Roman",
    24)).grid(row=0, column=1, columnspan=2)
# columnspan makes the text span two columns in our GUI
```

```
calculate_text = tk.Label(app, text="Convert ", font=text_font)
calculate_text.grid(row=1, column=1)

value_to_convert = tk.StringVar()
input_field = tk.Entry(app, borderwidth=2, relief="sunken",
                    textvariable=value_to_convert, font=text_font)
input_field.grid(row=1, column=2)

mpa_text = tk.Label(app, text=" MPa", font=text_font)
mpa_text.grid(row=1, column=3)

into_text = tk.Label(app, text='into', font=text_font)
into_text.grid(row=2, column=2)

calculated_value = tk.Label(app, text='', borderwidth=2,
                    relief="groove", width=15, font=text_font)
calculated_value.grid(row=3, column=2)

psi_text = tk.Label(app, text=' psi', font=text_font)
psi_text.grid(row=3, column=3)
```

Lastly, we need a button that will initiate the calculation. We will create a *convert_mpa_to_psi()* function later that will accept the number as a *str* and accept the *Label* field to update, and it will update the *Label* with the computed result. We will use the anonymous function syntax *lambda: example_function()* to re-compute each time the button is pressed. That way, you won't have to restart the GUI each time you want to change the value to convert. You can learn about lambda anonymous functions here [5]. This is the only time we will use them. We'll also add an empty *Label* to create an empty row between the result and the calculate button.

```
tk.Label(app, text="").grid(row=4, column=2)

convert_button = tk.Button(app, text="Convert",
            command=lambda: convert_mpa_to_psi(
            value_to_convert.get(), calculated_value),
            font=text_font)
```

Notice how we need to use *get()* on the *StringVar* to get what the user input is; otherwise, we would just pass along the *StringVar* instance with no *str* in it. When we want the GUI to run, we add:

```
app.mainloop()
```

Don't run the program just yet, though! We still need to create the *convert_mpa_to_psi()* function in the top of the program after the imports. The conversion factor from MPa to psi is 145.038. Our function will attempt to cast the passed-along *str* as a *float* and update the *Label* with the result. If the function is unable to cast the *str* as a *float*, it will update the *Label* with "NaN" meaning Not a Number.

```
def convert_mpa_to_psi(number: str, field: tk.Label):
    mpa_to_psi = 145.038
    try:
        toreturn = float(number.replace(',' '')) * mpa_to_psi
        field.configure(text='{:,.2f}'.format(toreturn),
                    relief="sunken")
    except ValueError:
        toreturn = 'NaN'
        field.configure(text='{}'.format(toreturn), relief="sunken")
```

Remember, `:,.2f` just means that the output will be a *float* that is rounded to 2 decimal places, and the comma separator will be used if the number is greater than 999. Phase 1 is finished! The full code should look like this:

```python
def convert_mpa_to_psi(number: str, field: tk.Label):
    mpa_to_psi = 145.038
    try:
        toreturn = float(number.replace(',', '')) * mpa_to_psi
        field.configure(text='{:,.2f}'.format(toreturn),
                        relief="sunken")
    except ValueError:
        toreturn = 'NaN'
        field.configure(text='{}'.format(toreturn), relief="sunken")

app = tk.Tk()
app.title('Unit Converter')

app.geometry('350x200')
text_font = ('Times New Roman', 14)

tk.Label(app, text="Converter", font=("Times New Roman",
                    24)).grid(row=0, column=1, columnspan=2)

calculate_text = tk.Label(app, text="Convert ", font=text_font)
calculute_text.grid(row=1, column=1)

value_to_convert = tk.StringVar()
input_field = tk.Entry(app, borderwidth=2, relief="sunken",
                       textvariable=value_to_convert, font=text_font)
input_field.grid(row=1, column=2)

mpa_text = tk.Label(app, text=" MPa", font=text_font)
mpa_text.grid(row=1, column=3)

into_text = tk.Label(app, text='into', font=text_font)
into_text.grid(row=2, column=2)

calculated_value = tk.Label(app, text='', borderwidth=2,
```

```
                    relief="groove", width=15, font=text_font)
calculated_value.grid(row=3, column=2)

psi_text = tk.Label(app, text=' psi', font=text_font)
psi_text.grid(row=3, column=3)

tk.Label(app, text="").grid(row=4, column=2)

convert_button = tk.Button(app, text="Convert", command=lambda:
                  convert_mpa_to_psi(value_to_convert.get(),
                  calculated_value), font=text_font)
convert_button.grid(row=5, column=2)
app.mainloop()
```

Hit run, and you should see something like this:

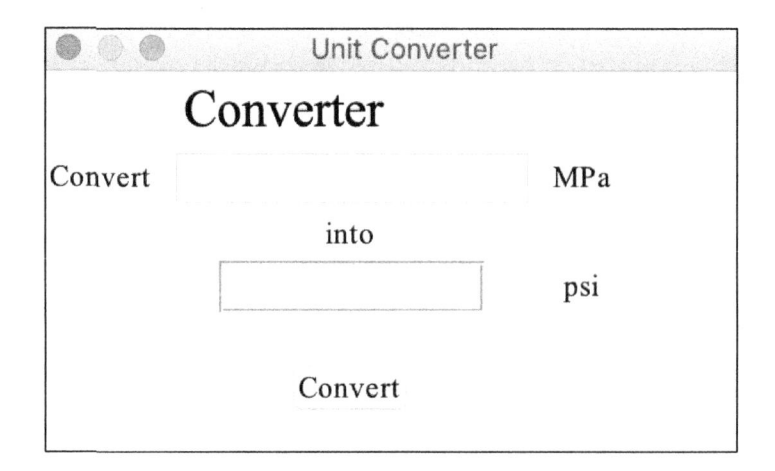

[Screenshot of Unit Converter GUI showing "Converter" title, "Convert" label, input field, "MPa" label, "into" text, output field, "psi" label, and "Convert" button]

Go ahead and type some numbers (and some not-numbers) into the field and see what happens! You may need to mess with your height and width settings to make the GUI fit your screen, and you may need to reset the Spyder graphics setting (Python > Preferences > iPython Console > Graphics) back to *Inline*.

Now we have a converter handy for converting MPa to psi. What if we want to convert between other units of pressure? Instead of hard-coding MPa and psi, we want to be able to give the user an *OptionMenu* of all of the available pressures from the Pint library. We'll create a *List* of all available pressure units, then make two *OptionMenus* to let the user pick which pressures to convert between. The GUI will now look like this:

	Column 1	Column 2	Column 3
Row 1		*Label* saying "Converter"	
Row 2	*Label* saying "Convert"	*Entry* field for inputting a *str*	*OptionMenu* with all pressures
Row 3		*Label* saying "into"	
Row 4		*Label* that will eventually have the computed result	*OptionMenu* with all pressures
Row 5		*Button* saying "Calculate"	

Let's look at how the Column 3 *OptionMenus* are created:

```
pressure_options = ["Pa", "Ba", "bar", "atm", "psi", "ksi", "mmHg",
                    "cmHg", "inch_Hg_60F", "inch_H2O_39F",
                    "inch_H2O_60F", "feet_H2O", "cm_H2O"]
first_units = tk.StringVar()
first_units.set(pressure_options[0])
units_one = tk.OptionMenu(app, first_units, *pressure_options)
units_one.grid(row=1, column=3)

second_units = tk.StringVar()
second_units.set(pressure_options[0])
units_two = tk.OptionMenu(app, second_units, *pressure_options)
units_two.grid(row=3, column=3)
```

133

Make sure the pressure *strs* in *pressure_options* are exactly as written in the Pint documentation; otherwise, the program won't be able to convert between the pressures. We set the default value of the *OptionMenu* as the first *str* in the *List*, but we could very easily make the default option random, or the first item and last item in the *Lists*, etc. We use the **pressure_options* notation to pass along all of the contents of *pressure_options*, not just the *List* itself.

We need to change our *Button* to pass along the additional variables *first_units* and *second_units* which represent the units selections that the user makes. We will also need to adjust our calculate function to recognize the units. For now, we need the new *calculate()* function to accept the user's input value, the unit that value is in, the unit that the user wants the original value to be converted to, and the field name of the widget that we will update:

```
convert_button = tk.Button(app, text="Convert", command=lambda:
                calculate(value_to_convert.get(), first_units.get(),
                second_units.get(), calculated_value))
```

We need to get Pint set up to accept our input. Pint has a *UnitRegistry()* which contains all of the units that Pint knows about. Within this registry is *Quantity* which represents units of quantity. We will create a converter function from *Quantity* that will help us associate a unit with a number.

```
ureg = pint.UnitRegistry()
converter = ureg.Quantity
```

Try the following in the Notebook:

```
>>> import pint
>>> ureg = pint.UnitRegistry()
>>> converter = ureg.Quantity
>>> converter(4.0, ureg.meter)
<Quantity(4.0, 'meter')>
>>> x = converter(4.0, ureg.meter)
>>> x.magnitude
4.0
>>> x.units
<Unit('meter')>
```

Fortunately, *ureg* also accepts a *str* input as *ureg('meter')* which is equal to *ureg.meter*. As an aside, Pint used to use a default Python function called *eval()* which attempts to convert the argument it receives into Python code and run it (*eval*uate it). This presents a security concern as someone could attempt to evaluate malicious code and have your computer execute the code. Pint previously used *eval()* to figure out what *str* input you passed *ureg*; since version 0.07, Pint no longer uses *eval()* due to the security concerns.

The other Pint function we will use is the *to()* function, which we can use to convert between units of the same dimension. If you still have the last code up, try:

```
>>> x.to('ft')
<Quantity(13.123359580052494, 'foot')>
>>> x.to('mph')
pint.errors.DimensionalityError: Cannot convert from 'meter'
([length]) to 'mile_per_hour' ([length] / [time])
```

Pint will only apply conversion factors where it makes sense: you cannot convert meters (distance) to mph (velocity) because the two cannot be converted. Pint also has support for metric prefixes; it understands how to go between kilometers (km) and meters (m), for example.

Let's get back to the program. Our new *calculate()* function needs to cast the user input as a *float*, use Pint to assign a unit to the *float*, use Pint to convert to the desired unit, and finally, update the *Label* widget with the new value.

```
def calculate(number: str, unit1: str, unit2: str,
              field: tk.Label) -> None:
    try:
        number = float(number.replace(',', ''))
        ureg = pint.UnitRegistry()
        converter = ureg.Quantity

        intial_value = converter(number, ureg(unit1))
        final_value = intial_value.to(unit2)
        field.configure(text='{:,.2f}'.format(
                        final_value.magnitude), relief="sunken")
    except ValueError:
        toreturn = 'NaN'
        field.configure(text='{}'.format(toreturn), relief="sunken")
```

The code should now look like this:

```
def calculate(number: str, unit1: str, unit2: str,
              field: tk.Label) -> None:
    try:
        number = float(number.replace(',', ''))
        ureg = pint.UnitRegistry()
        converter = ureg.Quantity

        intial_value = converter(number, ureg(unit1))
```

```python
        final_value = intial_value.to(unit2)
        field.configure(text='{:,.2f}'.format(

                        final_value.magnitude), relief="sunken")
    except ValueError:
        toreturn = 'NaN'
        field.configure(text='{}'.format(toreturn), relief="sunken")

app = tk.Tk()
app.title('Unit Converter')
app.geometry('350x200')
tk.Label(app, text="Converter", font=("Verdana", 24)).grid(row=0,
                                                    column=2)


calculate_text = tk.Label(app, text="Convert ")
calculate_text.grid(row=1, column=1)

value_to_convert = tk.StringVar()
input_field = tk.Entry(app, borderwidth=2, relief="sunken",
textvariable=value_to_convert, takefocus=True)
input_field.grid(row=1, column=2)

pressure_options = ["Pa", "Ba", "bar", "atm", "psi", "ksi", "mmHg",
                    "cmHg", "inch_Hg_60F", "inch_H2O_39F",
                    "inch_H2O_60F", "feet_H2O", "cm_H2O"]
first_units = tk.StringVar()
first_units.set(pressure_options[0])
units_one = tk.OptionMenu(app, first_units, *pressure_options)
units_one.grid(row=1, column=3)

into_text = tk.Label(app, text='into')
into_text.grid(row=2, column=2)

calculated_value = tk.Label(app, text='', borderwidth=2,
                            relief="groove", width=15)
calculated_value.grid(row=3, column=2)

second_units = tk.StringVar()
second_units.set(pressure_options[0])
units_two = tk.OptionMenu(app, second_units, *pressure_options)
```

```
units_two.grid(row=3, column=3)

tk.Label(app, text="").grid(row=4, column=2)

convert_button = tk.Button(app, text="Convert", command=lambda:
                calculate(value_to_convert.get(), first_units.get(),
                second_units.get(), calculated_value))
convert_button.grid(row=5, column=2)
app.mainloop()
```

And, when run, the program should look like this:

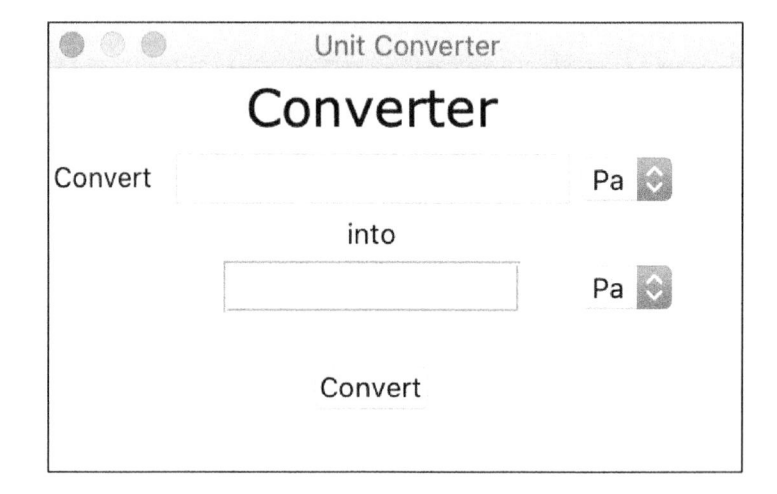

[Screenshot of Unit Converter GUI titled "Converter" with a "Convert" field, "into" label, a second field, Pa dropdowns on the right side, and a "Convert" button.]

Now, the user can jump between Pa and atm, bar and mmHg, and every combination in between. Importantly, we did not have to type in every single combination of pressure unit conversion factors to account for everything. However, with this layout, we are unable to use non-base units like converting from MPa to ksi (thousands of pounds per square inch) because we lost the ability to use any prefixes.

138

We'll make our final edit and change the *OptionMenus* to *Entry* fields and let Pint figure out the rest. We will get rid of the dropdown menus and let the user type in the units. The last iteration of our converter program will look like this:

	Column 1	Column 2	Column 3
Row 1		*Label* saying "Converter"	
Row 2	*Label* saying "Convert"	*Entry* field for inputting a *str*	*Entry* field for unit
Row 3		*Label* saying "into"	
Row 4		*Label* that will eventually have the computed result	*Entry* field for unit
Row 5		*Button* saying "Calculate"	

We need to edit the Column 3 widgets again. This time, we can get rid of any pressure *lists* and make the widget an open-ended *Entry* widget.

```
first_units = tk.StringVar()
first_units.set('Pa')
units_one = tk.Entry(app, borderwidth=2, relief="sunken",
                    textvariable=first_units, width=10)
units_one.grid(row=1, column=3)

second_units = tk.StringVar()
second_units.set('Pa')
units_two = tk.Entry(app, borderwidth=2, relief="sunken",
                    textvariable=second_units, width=10)
units_two.grid(row=3, column=3)
```

The *Button* command could be left alone, but because we are now dealing with user input on the units, let's *strip()* them before we call the calculate command (and while we're at it, we should probably *strip()* the user input value field, too). Remember, *strip()* will remove any leading or trailing whitespace like spaces, tabs, new lines, etc.

```
convert_button = tk.Button(app, text="Convert", command=lambda:
                    calculate(value_to_convert.get().strip(),
                    first_units.get().strip(),
                    second_units.get().strip(),
                    calculated_value))
```

There are two important errors we need to account for when trying to convert units: the *pint.errors.DimensionalityError* which means that you are attempting to convert between units that do not fundamentally represent the same quantity, and the *pint.errors.UndefinedUnitError* which means that the user input a unit that is not recognized. This undefined error is going to be particularly tricky for the proper capitalization of units like *MPa* vs *MPA* vs *mpa*. Only *MPa* will be recognized, while the other two will return an error. We could attempt to sanitize the input by looking for these common misspellings; however, we will just try to catch the unit error and hope the user can fix the mistake. The calculate function now looks like this:

```
def calculate(number: str, unit1: str, unit2: str,
            field: tk.Label) -> None:
    try:
        number = float(number.replace(',', ''))
        ureg = pint.UnitRegistry()
        converter = ureg.Quantity

        intial_value = converter(number, ureg(unit1))
        final_value = intial_value.to(unit2)
        field.configure(text='{:,.2f}'.format(
```

140

```
                    final_value.magnitude), relief="sunken")
except ValueError:
    toreturn = 'Input a number'
    field.configure(text='{}'.format(toreturn), relief="sunken")
except pint.errors.UndefinedUnitError:
    toreturn = 'Unrecognized unit'
    field.configure(text='{}'.format(toreturn), relief="sunken")
except pint.errors.DimensionalityError:
    toreturn = 'Check units'
    field.configure(text='{}'.format(toreturn), relief="sunken")
```

The full code should now read as:

```python
def calculate(number: str, unit1: str, unit2: str,
              field: tk.Label) -> None:
    try:
        number = float(number.replace(',', ''))
        ureg = pint.UnitRegistry()
        converter - ureg.Quantity

        intial_value = converter(number, ureg(unit1))
        final_value = intial_value.to(unit2)
        field.configure(text='{:,.2f}'.format(
                        final_value.magnitude), relief="sunken")
    except ValueError:
        toreturn = 'Input a number'
        field.configure(text='{}'.format(toreturn), relief="sunken")
    except pint.errors.UndefinedUnitError:
        toreturn = 'Unrecognized unit'
        field.configure(text='{}'.format(toreturn), relief="sunken")
    except pint.errors.DimensionalityError:
        toreturn = 'Check units'
        field.configure(text='{}'.format(toreturn), relief="sunken")

app = tk.Tk()
app.title('Unit Converter')

app.geometry('400x200')
```

```python
tk.Label(app, text="Converter", font=("Verdana", 24)).grid(row=0,
                                column=1, columnspan=3)

calculate_text = tk.Label(app, text="Convert ")
calculate_text.grid(row=1, column=1)

value_to_convert = tk.StringVar()
input_field = tk.Entry(app, borderwidth=2, relief="sunken",
                    textvariable=value_to_convert, width=15)
input_field.grid(row=1, column=2)

first_units = tk.StringVar()
first_units.set('Pa')
units_one = tk.Entry(app, borderwidth=2, relief="sunken",
                    textvariable=first_units, width=10)
units_one.grid(row=1, column=3)

into_text = tk.Label(app, text='into')
into_text.grid(row=2, column=2)

calculated_value = tk.Label(app, borderwidth=2, relief="groove",
                        width=15)
calculated_value.grid(row=3, column=2)

second_units = tk.StringVar()
second_units.set('Pa')
units_two = tk.Entry(app, borderwidth=2, relief="sunken",
                    textvariable=second_units, width=10)
units_two.grid(row=3, column=3)

tk.Label(app, text="").grid(row=4, column=2)

convert_button = tk.Button(app, text="Convert", command=lambda:
                    calculate(value_to_convert.get().strip(),
                    first_units.get().strip(),
                    second_units.get().strip(), calculated_value))
convert_button.grid(row=5, column=2)
app.mainloop()
```

Converter

Convert [ ] Pa

into

[ ] Pa

Convert

Because we took away the pressure *list*, this converter now works for all of the units that Pint knows about. We can convert MPa to ksi, mph to kph, ft to nanometers (nm), and all of the other units to our heart's content. Pint also provides functionality for defining new units [6]. You can keep an additional *list* that can be loaded and appended to the *UnitRegistry()* at import, or you can define a new unit in your Python program. Psf (pound-force per square foot) is another unit that is used for pressure, but Pint does not know about it. In the program, adding the psf unit would look like this:

```
ureg.define('pound_force_per_square_foot = force_pound / foot **2 = psf')
```

Defining any new units at the beginning of the program instead of trying to update the Pint settings is a better bet if you will be distributing the program at all. That way, you know the new unit will always be in the program, and you don't have to mess with any configuration settings or wonder if the new unit will be in Pint if you update Pint to the newest version.

Summary

Units are important, but they are also tedious to keep track of. Pint is a library that allows us to use Python to keep track of units. You can even use Pint and perform normal math operations and keep the units.

```
>>> ureg = pint.UnitRegistry()
>>> converter = ureg.Quantity
>>> x = converter(2.0, ureg.inches)
>>> x*12
>>> <Quantity(24.0, 'inch')>
```

GUIs provide a way for normal users to interact with a program; script editors and command line prompts are scary to most people, so the ability to create a GUI makes your program more appealing for the average consumer. It also allows you, as the author of the program, to be able to customize the program without having to write a bunch of *if* statements for exactly what you want. We were able to use the built-in library Tkinter to create a basic GUI to convert any unit into another unit in its dimension. Libraries like *cx_freeze*, *py2app*, or *py2exe* can be used to freeze Python programs and turn them into standalone computer applications (*.exe* for Windows, *.app* for Mac) that are distributable to people who do not have Python installed.

This program is still susceptible to bad user inputs like *Mpa* instead of *MPa*. There are several ways to try to make it more user-friendly: use a bunch of *replace()* statements, use a *dict* to convert to the proper capitalization, and more. See if you can implement one of these techniques to help the program be more user-friendly.

Full Code

```python
import tkinter as tk
import pint


# Third iteration of the GUI
def main():
    def calculate(number: str, unit1: str, unit2: str, field:
                tk.Label) -> None:
        try:
            number = float(number.replace(',', ''))
            ureg = pint.UnitRegistry()
            converter = ureg.Quantity

            intial_value = converter(number, ureg(unit1))
            final_value = intial_value.to(unit2)
            field.configure(text='{:,.2f}'.format(
                        final_value.magnitude), relief="sunken")
        except ValueError:
            toreturn = 'Input a number'
            field.configure(text='{}'.format(toreturn),
                        relief="sunken")
        except pint.errors.UndefinedUnitError:
            toreturn = 'Unrecognized unit'
            field.configure(text='{}'.format(toreturn),
                        relief="sunken")
        except pint.errors.DimensionalityError:
            toreturn = 'Check units'
            field.configure(text='{}'.format(toreturn),
                        relief="sunken")

    app = tk.Tk()
    app.title('Unit Converter')

    app.geometry('400x200')

    tk.Label(app, text="Converter", font=("Verdana",
```

```python
                            24)).grid(row=0, column=1, columnspan=3)

    calculate_text = tk.Label(app, text="Convert ")
    calculate_text.grid(row=1, column=1)
    value_to_convert = tk.StringVar()
    input_field = tk.Entry(app, borderwidth=2, relief="sunken",
                           textvariable=value_to_convert, width=15)
    input_field.grid(row=1, column=2)
    first_units = tk.StringVar()
    first_units.set('Pa')
    units_one = tk.Entry(app, borderwidth=2, relief="sunken",
                         textvariable=first_units, width=10)
    units_one.grid(row=1, column=3)
    into_text = tk.Label(app, text='into')
    into_text.grid(row=2, column=2)

    calculated_value = tk.Label(app, borderwidth=2, relief="groove",
                                width=15)
    calculated_value.grid(row=3, column=2)

    second_units = tk.StringVar()
    second_units.set('Pa')
    units_two = tk.Entry(app, borderwidth=2, relief="sunken",
                         textvariable=second_units, width=10)
    units_two.grid(row=3, column=3)
    tk.Label(app, text="").grid(row=4, column=2)

    convert_button = tk.Button(app, text="Convert", command=lambda:
                               calculate(value_to_convert.get().strip(),
first_units.get().strip(),
                               second_units.get().strip(),
                               calculated_value))
    convert_button.grid(row=5, column=2)
    app.mainloop()

if __name__ == '__main__':
    main()
```

View this code on GitHub at https://github.com/alexkenan/pymae/

Resources

[1] https://docs.python.org/3/library/tkinter.html

[2] https://effbot.org/tkinterbook/

[3] https://pint.readthedocs.io/en/latest/index.html

[4] https://github.com/hgrecco/pint/blob/master/pint/default_en.txt

[5] https://www.w3schools.com/python/python_lambda.asp

[6] https://pint.readthedocs.io/en/latest/defining.html

Chapter 7: Introduction to Web Scraping

Introduction

We used Requests earlier to scrape a website and do something with the results. That website functioned more like an API because we knew exactly what format the website would return. Having the data nicely formatted does not always happen. Sometimes, we have to find and format, clean, or otherwise manipulate the data ourselves. When we scrape a real website, we will need to be able to parse through HyperText Markup Language (HTML) to get to the information that we want. On any webpage that you visit, HTML and its sister Cascading Style Sheets (CSS) are hard at work to display the design of the particular webpage. The Engineering Toolbox website [1] contains technical data, material properties, chemical properties, economic information, drawing tools, and so much more for all types of engineers. We will make a Python program to grab information related to the material properties of different aluminum alloys from the Engineering Toolbox website. The program will put this information into an Excel spreadsheet so we can reference it later. BeautifulSoup4 is the preferred library for HTML parsing. The documentation for BeautifulSoup [2] is very good. Openpyxl is the preferred cross-platform library for manipulating Excel spreadsheets; Openpyxl's documentation can be found under footnote [3].

The program will need to go to the material properties page [4], convert the table into a nested *list* so the information is easily parse-able, and iterate through the info to put each value into a cell in an Excel spreadsheet.

Let's go through a rough crash course on HTML. It was the original way to build websites. HTML uses "tags" like *<a>text here</a>* to display different types of objects. The *<a>* tag creates a hyperlink, also called a link. In this case, the "text here" would show up as a blue, clickable link on a website like this. Below is a table of the common HTML tags:

HTML Tag	What it represents
<a>	Hyperlink
<div>, <p>	Paragraphs of text
<table>, <tr>, <td>, <tbody>, <thead>	Table, table row, and table cell, etc
<style>, <span>	Other types of paragraphs or text
<input>	A field to input data
<i>, <b>, <u>, <strike>, <sup>	Font styles like italics, bold, underline, strikethrough, superscript, respectively
<img>	Image

When information on a website is not nicely formatted, we have to dive into the HTML to get the information ourselves. This means that we will have to use Requests to get the text of the website, which means getting all of the source code. Then, we give the source code to BeautifulSoup to more easily parse through the code to get the information that we want. This allows us to treat the source code text as *lists*, so we can get only `<tr>` tags, `<a>` tags, etc.

Imports

```
import requests
from bs4 import BeautifulSoup
import openpyxl
```

All three of these libraries should come with Anaconda. If not, you will need to *pip install bs4 openpyxl requests*.

Body

The first function we want to create is one that will accept a website address (URL) as an argument and return the text (the code as *str*) of the website. This is a good, neutral function that can be included in any other program you write because it is agnostic to the URL.

```
def scrape_website(address: str) -> str:
    headers = {'user-agent': "Mozilla/5.0 (Macintosh; Intel Mac "\
                "OS X 10.15; rv:74.0) Gecko/20100101 Firefox/74.0"}
    # try to keep the headers on one line when writing the program
    r = requests.get(address, headers=headers)
    return r.text
```

Headers are what your browser sends along with its request to access a webpage. The user-agent defines what type of computer is making the request. Because requests without a user-agent are very obviously robots, it can be good practice to include your normal user-agent to show that you mean no harm. The easiest way to find your browser's user-agent is to type "What is my user agent" into a search engine. Requests will still work if you just *get()* the URL, but this type of request is the most likely to get blocked by a website since it is obviously coming from a bot. Some websites auto-block these requests to prevent different types of attacks. Having said that, don't add a user-agent to do malicious things to a website!

By itself, the *requests.Response* object contains the text of the website but also the status code, any json, history, encoding, URLs, etc. We just want the text, similar to what we used in Chapter 4 for plotting airfoils.

Next is the hard part: parsing the HTML. Constructing the "soup" is fairly simple: *soup = BeautifulSoup(response, features="lxml")*. This line uses the BeautifulSoup library to convert all of the text into a navigable BeautifulSoup object. *features* is an optional argument to tell BeautifulSoup which HTML parser to use. You may get an error if you pass *"lxml"*, and it is not installed. If you encounter this warning/error, just pass *soup = BeautifulSoup(response)* and pass the library that BeautifulSoup suggests.

Now we can navigate through the various tags to get the exact HTML snippet that we need. Though the website looks good when it is rendered on a browser for you, the code that makes it look pretty is a little more like spaghetti. A key element of web scraping is to dig through HTML code to find the "HTML coordinates" of the information that you are looking for. This requires looking through the website's source code to find HTML keywords or tags for the data that you are looking for.

Once you have found a tag to pull, the best way to find a tag you are looking for is by using the *soup.find()* method and supplying a *class_=* argument. Notice that the *class_* keyword has to have the underscore, because otherwise it would shadow the built-in *class* keyword, which is used to define and create new datatypes. The *find* method gives us a nested *list* of each level of tags, so you will usually have to use a couple of nested *for* loops to get to the data you want.

Our goal is to find all of the HTML that creates the Aluminum Properties table. The HTML on the right is what creates the webpage on the left. Specifically, what is highlighted on the right creates what is highlighted on the left.

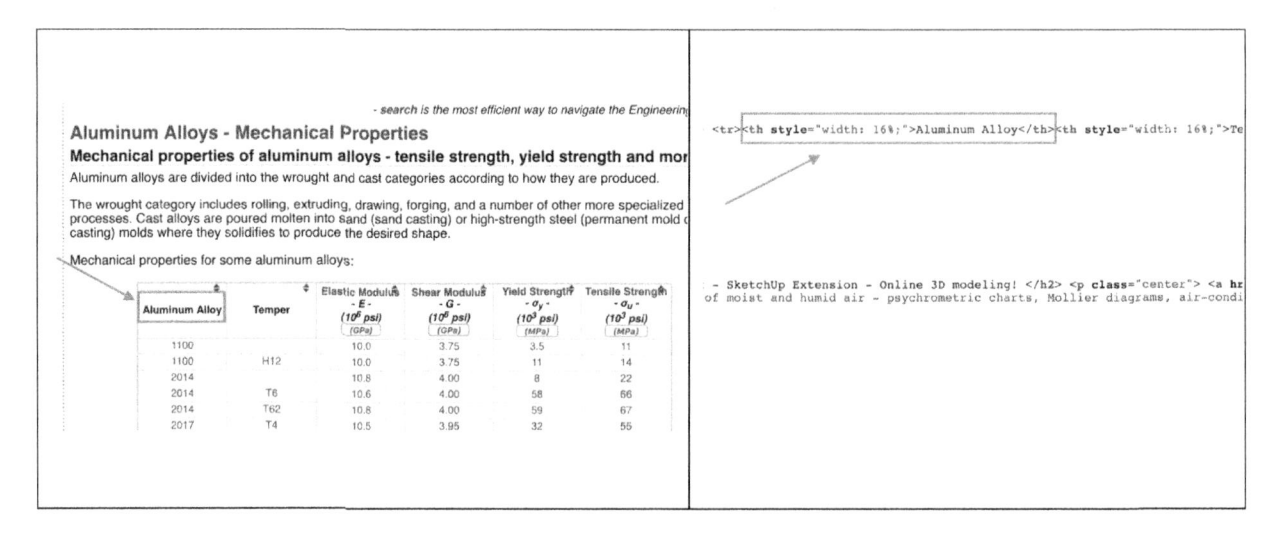

In order to find this source code, view the page source for that website by right-clicking on the website and selecting "View Page Source." Search for "Aluminum Alloy," which is the column header of the left-most column. The ninth instance of "Aluminum Alloy" is what we want: `<table class="large tablesorter"> <thead> <tr><th style="width: 16%;">Aluminum Alloy</th>`. This table class can also be found by right-clicking inside of the table on the webpage, clicking "Inspect Element," and looking to see if there is a table class or table ID. Finding a specific class name means that we can specifically call out the table that we want to scrape. We should be able to find the table with a class of *"large tablesorter"* in the soup. As a reminder, when you try to *find()* an HTML tag with a class property, the *class* keyword has to have an underscore after it. So, we will pass *class_="large tablesorter"* to the soup.

What comes next is a lot of guessing and checking to see how we can isolate the column headers and the subsequent rows by their HTML properties. By doing this, we get the exact "HTML coordinates" that we can use to always know how to find the material properties. The function *enumerate()* is a good troubleshooting tool to print out the indices and cells to help see if anything stands out.

```
table = soup.find('table', class_="large tablesorter")
for row in table:
    for index, tr in enumerate(row):
        print(index, tr)
```

This code printed a long list of tags to the console:

```
0
0
1 <tr><th style="width: 16%;">Aluminum Alloy</th> … <th></tr>
2
0
0
1 <tr> <td>1100</td> <td> </td> <td class="psi6">10.0</td> <td
class="psi6">3.75</td> <td class="psi6">3.5</td> <td
class="psi6">11</td> </tr>
2
3 <tr> <td>1100</td> <td>H12</td> <td class="psi6">10.0</td> <td
class="psi6">3.75</td> <td class="psi6">11</td> <td
class="psi6">14</td> </tr>
. . . . . .
78 <blah>
```

The second major line of code corresponds to this row in the aluminum properties table:

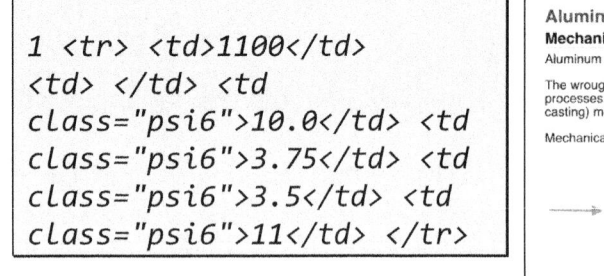

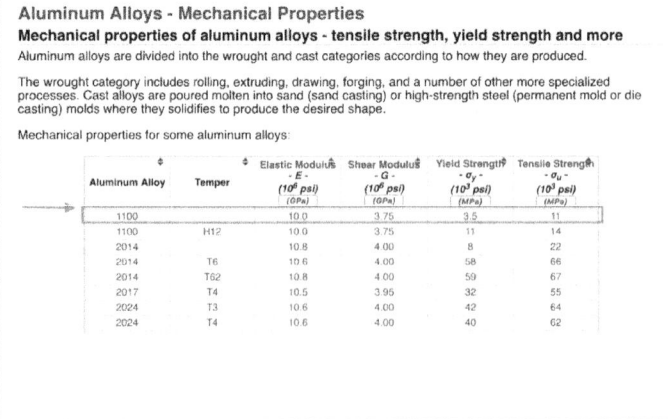

```
1 <tr> <td>1100</td>
<td> </td> <td
class="psi6">10.0</td> <td
class="psi6">3.75</td> <td
class="psi6">3.5</td> <td
class="psi6">11</td> </tr>
```

Since we can see the `<tr>` tag first (which means table row), that means that we can now start honing in on how we can filter out the blanks. Next step is to print the length of each *tr*:

154

```
table = soup.find('table', class_="large tablesorter")
for row in table:
    for tr in row:
        print(len(tr), tr)
```

Again, looking at the console:

```
1
1
6 <tr><th style="width: 16%;">Aluminum Alloy</th>...</tr>
1
1
1
13 <tr> <td>1100</td> <td> </td> <td class="psi6">10.0</td> <td
class="psi6">3.75</td> <td class="psi6">3.5</td> <td
class="psi6">11</td> </tr>
1
```

So, the header row has a length of 6, and the rows with the material properties have a length of 13. We can now iterate with another *for* loop to get rid of the *<tr>* tag and get to the *<td>* (table cell), but only on *<tr>* rows with length of 6 or 13.

```
table = soup.find('table', class_="large tablesorter")
for thead in table:
    for tr in thead:
        if len(tr) == 6:
            # this means we have the header row of the table
            internal_list = []
            for td in tr:
                if td:
                    internal_list.append(td.text)
            toreturn.append(internal_list)
        elif len(tr) == 13:
            # this means we have the material row
            internal_list = []
            for td in tr:
```

```
            if td and td.string != ' ':
                if str(td.string) == '\xa0':
                    value = 'None'
                else:
                    value = str(td.string)
                internal_list.append(value)
        toreturn.append(internal_list)
```

There are some blanks in the table, and each `<td>` tag is separated by a space `' '`, so we need to address those. The blanks in the table are a weird Unicode type character represented by `'\xa0'`. For the header row, the data scraping is easy. As long as the `td` is not empty, it has a header, so we grab it and add it to the running internal `list`. At the end of the `for` loop, we add the entire `list` to the big `list` `toreturn` that we will return. The actual material property rows are a little more complicated. We will filter out blanks and single space blanks, and further verify that the `td` isn't that weird Unicode character. At the end of the first `for` loop, we'll have a nested `list` of `lists`, where `toreturn[0]` is a `list` of the column headers, and the subsequent indices `toreturn[1:]` are `lists` of the material properties of each alloy. Here is what `toreturn` looks like:

```
[['Aluminum Alloy', 'Temper', 'Modulus - E - (106 psi) ',
    'Shear Modulus - G - (106 psi)',
    'Yield Strength - σy - (103 psi)',
  'Tensile Strength - σu -(103 psi)'],
 ['1100', 'None', '10.0', '3.75', '3.5', '11'],
 ['1100', 'H12', '10.0', '3.75', '11', '14'],
 ['2014', 'None', '10.8', '4.00', '8', '22'], ... ]
```

So, *toreturn[1] = ['1100', 'None', '10.0', '3.75', '3.5', '11']*, for example.

The full function that we created should look like this:

```python
def extract_properties_from_response(response: str) -> list:
    toreturn = []
    soup = BeautifulSoup(response, features="lxml")

    table = soup.find('table', class_="large tablesorter")
    for thead in table:
        for tr in thead:
            if len(tr) == 6:
                # this means we have the header row of the table
                internal_list = []
                for td in tr:
                    if td:
                        internal_list.append(td.text)
                toreturn.append(internal_list)
            elif len(tr) == 13:
                # this means we have the material row
                internal_list = []
                for td in tr:
                    if td and td.string != ' ':
                        if str(td.string) == '\xa0':
                            value = 'None'
                        else:
                            value = str(td.string)
                        internal_list.append(value)
                toreturn.append(internal_list)

    return toreturn
```

Lastly, we need to interact with Excel via the Openpyxl library. Below is sample code for interacting with a standard Excel workbook:

```
import openpyxl
workbook = openpyxl.Workbook()
worksheet = workbook.active  # or worksheet = workbook['sheet1']

# directly assign a cell a value
worksheet["A1"] = 'Start Here'
# or
worksheet.cell(row=1, column=1, value="Start Here")

# get the value of a cell
x = worksheet["A1"].value
# or
x = worksheet.cell(row=1, column=1).value
```

The functionality to assign a value to a cell by the exact cell name (A1) or by its row and column coordinates is a handy way to iterate through a *list* and dynamically assign the row and column with the *for* loop counter variables. With a double nested *for* loop, the outer *for* loop can keep track of the row number, and the inner *for* loop can keep track of the column number and keep assigning cell values until it runs out of indices. In code:

```
row_counter = 1
column_counter = 1
for row in info:
    # row represents each "line" or "row" of the table we scraped
    for cell in row:
        # cell represents each column or cell in the row
        worksheet.cell(row=row_counter, column=column_counter,
                       value=cell)
        # now increase the column counter variable by 1
        column_counter += 1
    """
    at this indentation, we need to reset the column variable to
    1 and increment the row variable by 1
```

```
"""
        column_counter = 1
        row_counter += 1
```

By making the code length agnostic, the number of rows in the table could increase and/or its number of columns could increase, and the program would still work. Openpyxl also has a *max_row* attribute that can be used to find the highest row number in the worksheet. This means you can use a *for* loop over a *range* instead of using the *while* loop to determine if the next cell is empty or not. Here is an example using that method:

```
max_row = worksheet.max_row
for i in range(1, max_row + 1):
    worksheet.cell(row=i, column=1).value = i
```

Back to our example: since all of the numbers are *str*, if we merely paste them into Excel, they will be text and not recognized as numbers. We can fix this by trying to cast each value as a *float*. We will also set the destination as the full path of the file. Let's call it *aluminum_properties.xlsx* and put it on the Desktop.

```
def create_spreadsheet(info: list) -> None:
    workbook = openpyxl.Workbook()
    worksheet = workbook.active
    worksheet.title = 'aluminum'
    destination = '/Users/Alex/Desktop/aluminum_properties.xlsx'

    row_counter = 1
    column_counter = 1
    for row in info:
        # row represents each "line" or "row" of the table we scraped
```

```
        for cell in row:
            # cell represents each column or cell in the row
            try:
                text = float(cell)
            except ValueError:
                text = cell
            worksheet.cell(row=row_counter, column=column_counter,
                            value=text)
            column_counter += 1
        column_counter = 1
        row_counter += 1

    workbook.save(filename=destination)
```

A word of warning for any Windows users: your destination path may need to be written like `r'C:\Users\Alex\Desktop\aluminum_properties.xlsx'`. If you don't do this, you may get the following error:

```
SyntaxError: (unicode error) 'unicodeescape' codec can't decode
bytes in position 2-3: truncated \UXXXXXXXX escape
```

This has to do with a peculiarity regarding \ being misinterpreted as an escape character. Because the Python interpreter can get confused in Windows file paths, putting the `r` in front of the `str` tells the interpreter to treat that `str` as a *raw* `str`.

With this function written, the main part of the program is only a handful of lines:

```
url = "https://www.engineeringtoolbox.com/"
url += 'properties-aluminum-pipe-d_1340.html'
# don't actually do this. Keep the URL on one line. This is the only
# way to get it to fit nicely onto the page
website_text = scrape_website(url)
results = extract_properties_from_response(website_text)
create_spreadsheet(results)
```

Each time the program is run, it will overwrite the current spreadsheet (if it exists) and replace its contents with the newly scraped data. Material properties of aluminum are not likely to change, but it is nice to know that the program will be able to adapt to any material additions or property additions since we used two *for* loops to keep putting info into cells as long as there is info in the argument *List*. The argument *List* is constructed by scraping the entire table from the website, so it should also be able to adapt to any additions.

Summary

We were able to demonstrate the powerful ability to go to a website, get information from it, and put that information into a spreadsheet. The program will be able to handle any column or row additions because of how we scrape the HTML. That does, however, make the program sensitive to any changes in the website's HTML or general layout. If the layout does change, you would have to re-find the proper HTML tag that identifies the information. This is part-and-parcel of depending on websites for information; occasionally, the style changes, and you have to make your program reflect the changes. This is why APIs are preferred: they typically will not change or will not change significantly.

The ability to interface with Excel with the *Openpyxl* library means you can decouple VBA from Excel and use Python to run any code you have. The days of seeing

IF(AND(OR(IFERROR(A5=A4, ""), A1 <> A6), B4 = TRUE), B3 = 1) may be over. Writing in Python is a better way to see readable logic, and now you know how to scrape websites and use Excel to display the results.

Full Code

```
import requests
from bs4 import BeautifulSoup
import openpyxl

def scrape_website(address: str) -> str:
    headers = {'user-agent': "Mozilla/5.0 (Macintosh; Intel Mac OS \
            X 10.15; rv:74.0) Gecko/20100101 Firefox/74.0"}
    r = requests.get(address, headers=headers)
    return r.text

def extract_properties_from_response(response: str) -> list:
    toreturn = []
    soup = BeautifulSoup(response, features="lxml")
    table = soup.find('table', class_="large tablesorter")
    for thead in table:
        for tr in thead:
            if len(tr) == 6:
                # this means we have the header row of the table
                internal_list = []
                for td in tr:
                    if td:
                        internal_list.append(td.text)
                toreturn.append(internal_list)
            elif len(tr) == 13:
                # this means we have the material row
                internal_list = []
                for td in tr:
                    if td and td.string != ' ':
                        if str(td.string) == '\xa0':
                            value = 'None'
                        else:
```

```python
                        value = str(td.string)
                        internal_list.append(value)
                toreturn.append(internal_list)

    return toreturn

def create_spreadsheet(info: list) -> None:
    workbook = openpyxl.Workbook()
    worksheet = workbook.active
    worksheet.title = 'aluminum'
    destination = '/Users/Alex/Desktop/aluminum_properties.xlsx'
    row_counter = 1
    column_counter = 1
    for row in info:
        # row represents each "line" or "row" of the table we scraped
        for cell in row:
            # cell represents each column or cell in the row
            try:
                text = float(cell)
            except ValueError:
                text = cell
            workshee.cell(row=row_counter, column=column_counter,
                          value=text)
            column_counter += 1
        column_counter = 1
        row_counter += 1

    workbook.save(filename=destination)

url = "https://www.engineeringtoolbox.com/"
url += 'properties-aluminum-pipe_d_1340.html'
# don't actually do this. Keep the URL on one line. This is the only
# way to get it to fit nicely onto the page
website_text = scrape_website(url)
results = extract_properties_from_response(website_text)
create_spreadsheet(results)
```

View this code on GitHub at https://github.com/alexkenan/pymae/

Resources

[1] https://www.engineeringtoolbox.com/

[2] https://www.crummy.com/software/BeautifulSoup/bs4/doc/

[3] https://openpyxl.readthedocs.io/en/stable/

[4] https://www.engineeringtoolbox.com/properties-aluminum-pipe-d_1340.html

Chapter 8: Modeling Camera Shutter Effect

Introduction

If you have ever seen a video of an aircraft propeller or a helicopter rotor, you may have noticed that the spinning rotor looks strange. Rather than looking like a filled-in circle, sometimes the propeller does not move at all, and sometimes the propeller looks like a saggy banana or wilted flower. Image search "propeller camera effect" if you need a better visual.

The shutter effect happens because the camera shutter is essentially a line moving down the screen recording whatever it touches. The propeller is spinning faster than the line is moving, so the propeller ends up looking disfigured because of the speed mismatch. We can use Tkinter, the Python Imaging Library (PIL or Pillow), and our trusty friend Numpy (the numerical Python library) to model a spinning propeller and a panning shutter line to display what the camera sees in real-time.

Our program will use Numpy to create arrays for the panning line and spinning propeller, use Pillow to convert an array to an image, and use Tkinter to display the image. We will write the program so that we can input a propeller of between 2 and 8 blades to see how the shutter effect changes with blade count. This program is based off of a program created by GitHub user *dafarry*, and the original program can be found here [1] This program requires very advanced usage of Numpy with complex geometric planes. Explanations will be attempted and provided where possible, but it may be necessary to accept incomplete understanding at times. Such is life! This chapter is designed to show applications of advanced Python usage.

Imports

```
import tkinter as tk
import numpy as np
from PIL import Image, ImageTk
```

We have not worked with PIL yet, so run *pip install Pillow*. Pillow was the underlying writer that made our gifs in Chapter 5.

Body

This program will use conic sections to compute the ellipses that will represent the propellers. Here is a very high-level mathematical explanation: the absolute value of a complex plane in the third dimension is a cone. Slicing the cone in planes parallel to the x-y plane results in a circle. Anything from the circle down to the Z = 0 plane is a disk. Adding multiple cones offset from the origin results in a solid ellipse, and moving the location of the ellipse around the origin X amount of times results in X number of blades in a propeller. Conic sections continue to haunt you, and this time, they're *complex*! This program glosses over some difficult topics like representing a two-dimensional array with one variable with Numpy, creating geometric shapes from simple expressions of complex numbers, and using addition/multiplication to translate the geometric shapes.

First, we need to get Tkinter configured. We will create a *Tk* instance that is the base Tkinter instance, get the "stage" ready that is where the image goes, and set the stage height and number of propeller blades.

```
root = tk.Tk()
stage = tk.Label(root)
stage.pack()
height = 300
num_propeller_blades = 4
```

The next snippet creates a meshgrid that returns a three dimensional array; this array helps us to create the 3D plane that will create the spinning propeller. We also establish the bent propeller image from the complex plane that will start out blank (*zeros*) and eventually be filled in when the line overlaps the spinning propeller.

```
x, y = np.ogrid[-height/2: height/2, -height/2: height/2]
plane = x - 1j * y
bentprop = np.zeros_like(plane, dtype=np.bool)
```

You can learn more about *ogrid()* here [2]. The - 1 j means the propeller will spin clockwise; change it to + 1 j for counter-clockwise rotation. For each "row" in our frame height (300 pixels), we need to sweep the shutter down one pixel, create the propeller blades, rotate the propeller blades, and "remember" where the shutter line touched the spinning propeller blade.

```
for frame in range(height):
    backgnd = np.zeros_like(plane, dtype=np.uint8)
    propeller = np.zeros_like(plane, dtype=np.bool)
    angle = 2 * np.pi * (frame / height)
    for blade in range(num_propeller_blades):
        phase = np.exp(1j * (angle + blade * np.pi /
                    (num_propeller_blades / 2)))
        ellipse = abs(plane - 0.48 * height * phase) + abs(plane)
        propeller |= ellipse < 0.49 * height
```

Now that we have all of the image data, we need to convert the arrays into an actual image using Pillow, and then update the stage with the new image.

```
bentprop[frame] = propeller[frame]
shutter_pan = [f for f in range(frame, min(frame + 3, height - 3))]
colors = ("white", "black", "red", "blue")
rgbcolors = np.array(list(map(root.winfo_rgb, colors))) / 256
composite = np.maximum.reduce((backgnd, propeller * 1,
                    bentprop * 2))
composite[shutter_pan] = 3
rgb = rgbcolors.astype(np.uint8)[composite]
image = ImageTk.PhotoImage(Image.fromarray(rgb, mode="RGB"))
stage.config(image=image)
root.update()
```

This snippet needs to occur for each frame update too, so it needs to be included in the first *for* loop:

```
for frame in range(height):
    backgnd = np.zeros_like(plane, dtype=np.uint8)
    propeller = np.zeros_like(plane, dtype=np.bool)
    angle = 2 * np.pi * (frame / height)
    for blade in range(num_propeller_blades):
        phase = np.exp(1j * (angle + blade * np.pi /
                    (num_propeller_blades / 2)))
        ellipse = abs(plane - 0.48 * height * phase) + abs(plane)
```

```
        propeller |= ellipse < 0.49 * height

    bentprop[frame] = propeller[frame]
    shutter_pan = [f for f in range(frame, min(frame + 3,
                                    height - 3))]
    colors = ("white", "black", "red", "blue")
    rgbcolors = np.array(list(map(root.winfo_rgb, colors))) / 256
    composite = np.maximum.reduce((backgnd, propeller * 1,
                                   bentprop * 2))
    composite[shutter_pan] = 3
    rgb = rgbcolors.astype(np.uint8)[composite]
    image = ImageTk.PhotoImage(Image.fromarray(rgb, mode="RGB"))
    stage.config(image=image)
    root.update()
```

Lastly, we need to make sure the GUI snaps to the front, and we'll give the new window a nice title.

```
root.lift()
root.attributes('-topmost', True)
root.after_idle(root.attributes, '-topmost', False)
root.title('{} blade propeller camera effect'.format(
                                num_propeller_blades))
root.mainloop()
```

Here is the full code:

```python
root = tk.Tk()
stage = tk.Label(root)
stage.pack()
height = 300
num_propeller_blades = 4
x, y = np.ogrid[-height/2: height/2, -height/2: height/2]
plane = x - 1j * y
bentprop = np.zeros_like(plane, dtype=np.bool)

for frame in range(height):
    backgnd = np.zeros_like(plane, dtype=np.uint8)
    propeller = np.zeros_like(plane, dtype=np.bool)
    angle = 2 * np.pi * (frame / height)
    for blade in range(num_propeller_blades):
        phase = np.exp(1j * (angle + blade * np.pi /
                        (num_propeller_blades / 2)))
        ellipse = abs(plane - 0.48 * height * phase) + abs(plane)
        propeller |= ellipse < 0.49 * height

    bentprop[frame] = propeller[frame]
    shutter_pan = [f for f in range(frame, min(frame + 3,
                                    height - 3))]
    colors = ("white", "black", "red", "blue")
    rgbcolors = np.array(list(map(root.winfo_rgb, colors))) / 256
    composite = np.maximum.reduce((backgnd, propeller * 1,
                                    bentprop * 2))
    composite[shutter_pan] = 3
    rgb = rgbcolors.astype(np.uint8)[composite]
    image = ImageTk.PhotoImage(Image.fromarray(rgb, mode="RGB"))
    stage.config(image=image)
    root.update()

root.lift()
root.attributes('-topmost', True)
root.after_idle(root.attributes, '-topmost', False)
root.title('{} blade propeller camera
            effect'.format(num_propeller_blades))
root.mainloop()
```

The code should produce a moving image, and when complete, should look something like this:

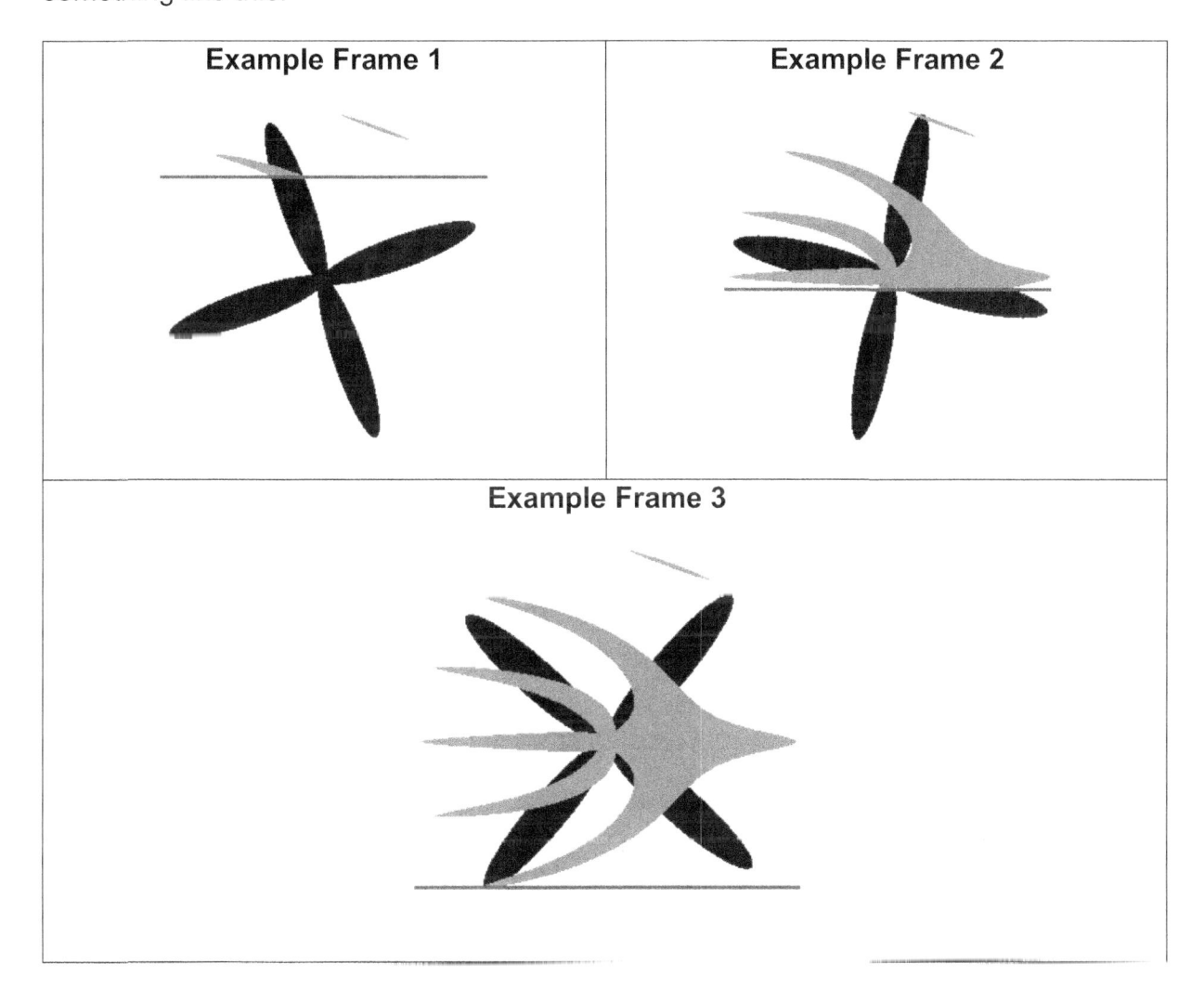

Example Frame 1

Example Frame 2

Example Frame 3

The red (light gray if you're reading this in black and white) represents every time that a propeller blade interacted with the blue line (a slightly darker light gray in black and white). If you took a picture of a spinning propeller, the photograph would look like the red bird/blob and not the 4-bladed propeller. You can vary the number of blades and the rotation direction to see how the shutter image changes. The propeller starts getting pretty fat after 7 blades, and you'll be hard-pressed to find a propeller that has more than 6 blades. This graphic easily explains why the propellers look like they're made of rubber when on camera: the shutter only picks up some of the blades, some of the time, so the resulting image is not a faithful representation of the actual propeller rotation.

Summary

Using Numpy gives Python the full array implementation that it is so frequently accused of lacking. With Numpy, Pillow, and Tkinter, we were able to loop over a frame height, create individual frame arrays, and display them in a standalone GUI. The implementation requires complex conic section knowledge to make the arrays, but this program easily could have used shapes to accomplish the same result.

This chapter shows how you can perform advanced calculation, modeling, and simulation in Python. Though it glosses over some of the complex subjects, Numpy and Tkinter can be used to demonstrate advanced concepts in otherwise simple programs.

Full Code

```python
import tkinter as tk
import numpy as np
from PIL import Image, ImageTk

root = tk.Tk()
stage = tk.Label(root)
stage.pack()
height = 300
num_propeller_blades = 4
x, y = np.ogrid[-height/2: height/2, -height/2: height/2]
plane = x - 1j * y
bentprop = np.zeros_like(plane, dtype=np.bool)

for frame in range(height):
    backgnd = np.zeros_like(plane, dtype=np.uint8)
    propeller = np.zeros_like(plane, dtype=np.bool)
    angle = 2 * np.pi * (frame / height)
    for blade in range(num_propeller_blades):
        phase = np.exp(1j * (angle + blade * np.pi /
                    (num_propeller_blades / 2)))
        ellipse = abs(plane - 0.48 * height * phase) + abs(plane)
        propeller |= ellipse < 0.49 * height

    bentprop[frame] = propeller[frame]
    shutter_pan = [f for f in range(frame, min(frame + 3,
                                    height - 3))]
    colors = ("white", "black", "red", "blue")
    rgbcolors = np.array(list(map(root.winfo_rgb, colors))) / 256
    composite = np.maximum.reduce((backgnd, propeller * 1,
                                    bentprop * 2))
    composite[shutter_pan] = 3
    rgb = rgbcolors.astype(np.uint8)[composite]
    image = ImageTk.PhotoImage(Image.fromarray(rgb, mode="RGB"))
    stage.config(image=image)
    root.update()

root.lift()
```

```
root.attributes('-topmost', True)
root.after_idle(root.attributes, '-topmost', False)
root.title('{} blade propeller camera effect'.format(
                                        num_propeller_blades))
root.mainloop()
```

View this code on GitHub at https://github.com/alexkenan/pymae/

Resources

[1] https://github.com/dafarry/rolling-shutter-effect-python

[2] https://towardsdatascience.com/the-little-known-ogrid-function-in-numpy-19ead3bdae40

Chapter 9: Writing Reports with Pweave

Introduction

It is one thing to have a good program; it is another to be able to share it with others. MATLAB has a built-in share feature that takes the code and turns it into a shareable file format like HTML webpage, PDF, Markdown, and others. Natively, Python does not have this capability; however, we will use the Pweave third-party library to generate HTML from the Python program which can be saved as a PDF for distribution. This technique allows us to open the report with a web browser and use the native "Save to PDF" functionality for distribution.

Pweave is a report generator that can run Python code, capture the input/output, and display it in several different formats. We'll use the program developed to graphically show Max Q in Chapter 2 and turn it into a shareable PDF. The documentation for Pweave can be found here [1], though it is spotty at best. By adding a few comment lines that simulate Markdown [2] formatting, Python programs can be turned into reports that show the input code, display the results, and hide any intermediate sections that detract from the presentation. We will make a few edits to the existing Max Q program, then we will use a command terminal to generate an HTML file that can be saved as a PDF and distributed as necessary.

Imports

```
import numpy as np
import matplotlib.pyplot as plt
```

The imports for the program are the same as in Chapter 3; however, you will need to install the Pweave library with `pip install pweave`.

Body

Pweave uses Markdown to provide basic formatting like bolding text, italicizing text, and providing different header sizes. Here is a rundown on Markdown:

Markdown	Result
`# Text`	**Heading 1 (like a title font)**
`## Text`	**Heading 2 (subtitle font)**
`### Text`	**Heading 3**
`**Text**`	**Text**
`*Text*`	*Text*

In Python, a line with formatting needs to first have the single-line comment character #, then a single quote ', then the Markdown formatting.

```
#' # This is equivalent to Heading 1 font
#' ## This is Heading 2
#' ### Heading 3
#' #### etc
#' **This is bold**
#' *This is italics*
```

There is also a marker *#+echo=False* to hide the printout of anything after that marker until the next section or end of file is reached. For example, if you wanted to hide the imports section:

```
#' ## Imports
#+echo=False
......
#' ## Calculations
```

Anything between the *#+echo=False* and *#' ## Calculations* would be hidden from the new version.

The Max Q program needs to be adjusted slightly to have all of the plotting at the end; as a result, the max values need to be broken out into different variables so that the last max value does not overwrite the others. The headers are also added before the imports, before the equations, before the calculation, and before the plotting.

```
#' # Dynamic Pressure for a Rocket Launch
#' ** Author**: Alex Kenan

#' ## Imports
import numpy as np
import matplotlib.pyplot as plt

#' ##Equations
```

```python
def density(height: float) -> float:
    if height < 36152.0:
        temp = 59 - 0.00356 * height
        p = 2116 * ((temp + 459.7)/518.6)**5.256
    elif 36152 <= height < 82345:
        p = 473.1*np.exp(1.73 - 0.000048*height)
    else:
        temp = -205.05 + 0.00164 * height
        p = 51.97*((temp + 459.7)/389.98)**-11.388

    rho = p/(1718*(temp + 459.7))
    return rho

def velocity(time: float, acceleration: float) -> float:
    return acceleration*time

def altitude(time: float, acceleration: float) -> float:
    return 0.5*acceleration*time**2

#' ## Calculations
y_values = []
x_values = np.arange(0.0, 550.0, 0.5)
for elapsed_time in x_values:
    accel = 51.764705882
    alt = altitude(elapsed_time, accel)
    # Dynamic pressure q = 0.5*rho*V^2 = 0.5*density*velocity^2
    q = 0.5 * density(alt) * velocity(elapsed_time, accel) ** 2
    y_values.append(q)

max_val = max(y_values)
ind = y_values.index(max_val)

# Now let's make acceleration = 32.2 ft/s^2 (1g)
y2_values = []
for elapsed_time in x_values:
    accel = 32.2
    alt = altitude(elapsed_time, accel)
    q = 0.5 * density(alt) * velocity(elapsed_time, accel) ** 2
    y2_values.append(q)
```

```python
max_val2 = max(y2_values)
ind2 = y2_values.index(max_val2)

# Now let's make acceleration = 20.0 ft/s^2 (1g)
y3_values = []
for elapsed_time in x_values:
    accel = 20.0
    alt = altitude(elapsed_time, accel)
    q = 0.5 * density(alt) * velocity(elapsed_time, accel) ** 2
    y3_values.append(q)

max_val3 = max(y3_values)
ind3 = y3_values.index(max_val3)

#' ## Plotting
plt.style.use('bmh')

# Plot the first line and Max Q
plt.plot(x_values, y_values, 'b-',
        label=r"a = 51.76 $\frac{ft}{s^2}$")

plt.annotate('{:,.0f} psf @ {} s'.format(max_val, x_values[ind]),
            xy=(x_values[ind] + 2, max_val),
            xytext=(x_values[ind] + 15, max_val + 15),
            arrowprops=dict(facecolor='blue', shrink=0.05))

plt.plot(x_values[ind], max_val, 'rx')

# Plot the second line and Max Q
plt.plot(x_values, y2_values, 'k-',
        label=r"a = 32.2 $\frac{ft}{s^2}$")

plt.annotate('{:.0f} psf @ {} s'.format(max_val2, x_values[ind2]),
            xy=(x_values[ind2] + 3, max_val2),
            xytext=(x_values[ind2] + 15, max_val2 + 15),
            arrowprops=dict(facecolor='black', shrink=0.05))

plt.plot(x_values[ind2], max_val2, 'rx')
```

```
# Plot the third line and Max Q

plt.plot(x_values, y3_values, 'g-',
         label=r"a = 20.0 $\frac{ft}{s^2}$")

plt.annotate('{:.0f} psf @ {} s'.format(max_val3, x_values[ind3]),
             xy=(x_values[ind3] + 3, max_val3),
             xytext=(x_values[ind3] + 15, max_val3 + 15),
             arrowprops=dict(facecolor='green', shrink=0.05))

plt.plot(x_values[ind3], max_val3, 'rx')

# Plot cleanup
plt.xlim(0, 150)
plt.xlabel('Time (s)')
plt.ylabel('Pressure (psf)')
plt.title('Dynamic pressure as a function of time')
plt.legend()
plt.show()
```

When using Pweave from the command line, we really only have one argument to supply, which is the desired output format. Here are all of the output formats that Pweave can generate:

- tex
- texminted
- texpweave
- texpygments
- rst
- pandoc
- sphinx
- html
- md2html
- softcover
- pandoc2latex
- pandoc2html

- markdown
- leanpub
- notebook

Using a CMD/Terminal window, *pweave -l* (lower case L) will show all of the available formats and explanations behind the formatting. We put Markdown in our Python program, so we want *md2html*. In CMD/Terminal, you need to navigate to the folder that contains the program you will be converting to HTML. For me, the path is */Users/Alex/Documents/Python3/book/chapter9/*. On Mac and in Terminal, this command looks like:

```
cd /Users/Alex/Documents/Python3/book/chapter9/
```

On Windows and in CMD/Anaconda Prompt:

```
cd C:\Users\Alex\Documents\Python3\book\chapter9\
```

Make sure that you use the correct type of slash for your system. / for Mac, \ for Windows. The program we will be Pweaving is called *chap9.py*; make sure the program you type is the correct program name. Use the following command to Pweave your program:

```
pweave -f md2html chap9.py
```

There should now be a file in the same folder called *(filename).html*. For me, it is *chap9.html*. Open it with a web browser, and you should see your report in a nice, shareable format! The best way to get this HTML file to a PDF is to try to print it and save to PDF. If we wanted just the plot and no code to show, we could get rid of the headers in the comments and add *+echo=False* statements like this:

```
#' # Dynamic Pressure for a Rocket Launch
#' ** Author**: Alex Kenan
#+echo=False
# ## Imports
#+echo=False
......

# ##Equations
#+echo=False
......

# ## Calculations
#+echo=False
......

# ## Plotting
......
```

You will have to take away the ' in front of the headers you do not want to show. Here is the result:

Dynamic Pressure for a Rocket Launch

Author: Alex Kenan

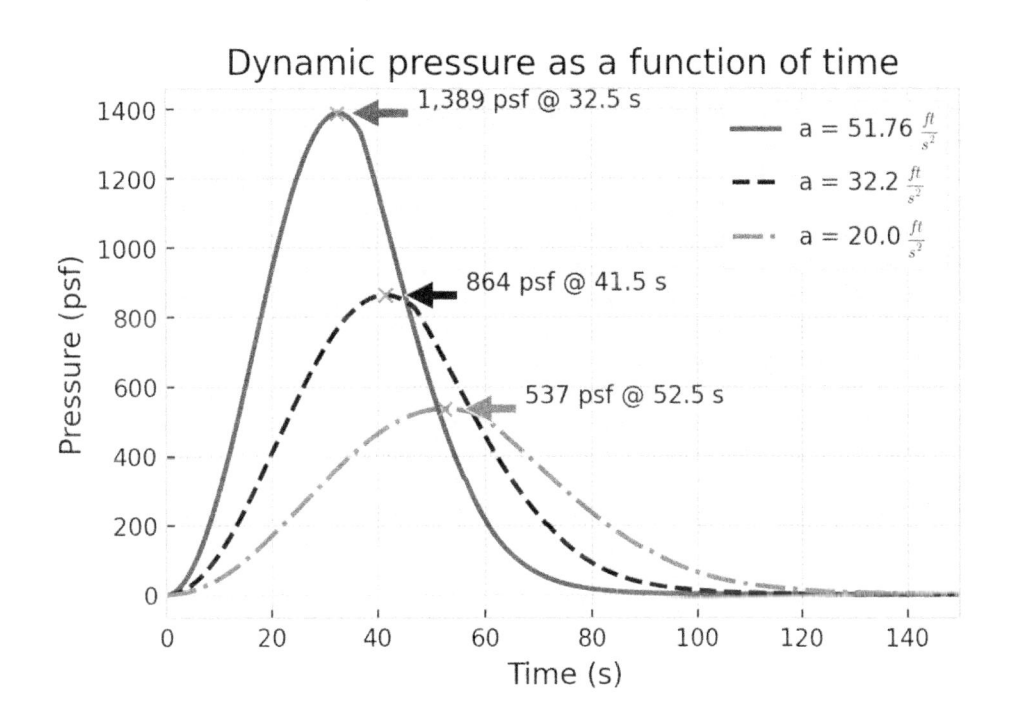

Dynamic pressure as a function of time

[Figure: Plot showing dynamic pressure (psf) vs Time (s) for three acceleration values. Peak values labeled: 1,389 psf @ 32.5 s for $a = 51.76 \frac{ft}{s^2}$; 864 psf @ 41.5 s for $a = 32.2 \frac{ft}{s^2}$; 537 psf @ 52.5 s for $a = 20.0 \frac{ft}{s^2}$.]

That wouldn't fare so well as a standalone report, but it would be a cool graphic to distribute or put into a slideshow. (The linestyle has also been adjusted to display better in black and white print.) If you just needed the graphic, the most practical way to obtain it would be to screenshot it after running the program, before we do any Pweave nonsense.

Summary

Pweave is the final tool in the toolbox to be able to broadcast and publish your Python programs to a wider audience that wants to see the results, not necessarily the program. Pweave provides the flexibility to show or hide as much code as wanted while still displaying the end results. The Markdown formatting does not detract from the Python code, and the program itself will still run even with the commented Markdown formatting in place. Pweave is designed to work with Numpy, Scipy, Matplotlib, and default print and display behavior, but it should be able to handle just about anything else thrown at it.

Full Code

```
#+echo=False
#!/usr/bin/env python3
#' # Dynamic Pressure for a Rocket Launch
#' ** Author**: Alex Kenan

#' ## Imports
import numpy as np
import matplotlib.pyplot as plt

#' ##Equations
def density(height: float) -> float:
    if height < 36152.0:
        temp = 59 - 0.00356 * height
        p = 2116 * ((temp + 459.7)/518.6)**5.256
    elif 36152 <= height < 82345:
        p = 473.1*np.exp(1.73 - 0.000048*height)
    else:
        temp = -205.05 + 0.00164 * height
        p = 51.97*((temp + 459.7)/389.98)**-11.388

    rho = p/(1718*(temp + 459.7))
    return rho

def velocity(time: float, acceleration: float) -> float:
    return acceleration*time

def altitude(time: float, acceleration: float) -> float:
```

```python
    return 0.5*acceleration*time**2

#' ## Calculations
y_values = []
x_values = np.arange(0.0, 550.0, 0.5)
for elapsed_time in x_values:
    accel = 51.764705882
    alt = altitude(elapsed_time, accel)
    # Dynamic pressure q = 0.5*rho*V^2 = 0.5*density*velocity^2
    q = 0.5 * density(alt) * velocity(elapsed_time, accel) ** 2
    y_values.append(q)

max_val = max(y_values)
ind = y_values.index(max_val)

# Now let's make acceleration = 32.2 ft/s^2 (1g)
y2_values = []
for elapsed_time in x_values:
    accel = 32.2
    alt = altitude(elapsed_time, accel)
    q = 0.5 * density(alt) * velocity(elapsed_time, accel) ** 2
    y2_values.append(q)

max_val2 = max(y2_values)
ind2 = y2_values.index(max_val2)

# Now let's make acceleration = 20.0 ft/s^2 (1g)
y3_values = []
for elapsed_time in x_values:
    accel = 20.0
    alt = altitude(elapsed_time, accel)
    q = 0.5 * density(alt) * velocity(elapsed_time, accel) ** 2
    y3_values.append(q)

max_val3 = max(y3_values)
ind3 = y3_values.index(max_val3)

#' ## Plotting
plt.style.use('bmh')
```

```python
# Plot the first line and Max Q
plt.plot(x_values, y_values, 'b-',
        label=r"a = 51.76 $\frac{ft}{s^2}$")

plt.annotate('{:,.0f} psf @ {} s'.format(max_val, x_values[ind]),
            xy=(x_values[ind] + 2, max_val),
            xytext=(x_values[ind] + 15, max_val + 15),
            arrowprops=dict(facecolor='blue', shrink=0.05))

plt.plot(x_values[ind], max_val, 'rx')

# Plot the second line and Max Q
plt.plot(x_values, y2_values, 'k-',
        label=r"a = 32.2 $\frac{ft}{s^2}$")

plt.annotate('{:.0f} psf @ {} s'.format(max_val2, x_values[ind2]),
            xy=(x_values[ind2] + 3, max_val2),
            xytext=(x_values[ind2] + 15, max_val2 + 15),
            arrowprops=dict(facecolor='black', shrink=0.05))

plt.plot(x_values[ind2], max_val2, 'rx')

# Plot the third line and Max Q
plt.plot(x_values, y3_values, 'g-',
        label=r"a = 20.0 $\frac{ft}{s^2}$")
plt.annotate('{:.0f} psf @ {} s'.format(max_val3, x_values[ind3]),
            xy=(x_values[ind3] + 3, max_val3),
            xytext=(x_values[ind3] + 15, max_val3 + 15),
            arrowprops=dict(facecolor='green', shrink=0.05))
plt.plot(x_values[ind3], max_val3, 'rx')

# Plot cleanup
plt.xlim(0, 150)
plt.xlabel(r'Time (s)')
plt.ylabel('Pressure (psf)')
plt.title(r'Dynamic pressure as a function of time')
plt.legend()
plt.show()
```

Then navigate to the folder that contains the program with Anaconda Prompt (Windows) or Terminal (Mac/Linux).

Windows	Mac
`cd C:\path\to\folder\`	`cd /path/to/folder/`

Use the following command to generate the HTML file:

```
pweave -f md2html {file_name}.py
```

View this code on GitHub at https://github.com/alexkenan/pymae/

Resources

[1] http://mpastell.com/pweave/index.html

[2] https://en.wikipedia.org/wiki/Markdown

The End

Well, you have reached the end of the book. How was it? I decided to write this book to address the gap that I saw between mechanical/aerospace engineering and Python. Hopefully, the use of engineering examples showed you how Python can help you in your everyday life. I could not fit everything I wanted into this book. If you have any ideas for inclusions in a 2nd edition, use the below methods to get in contact.

I decided to license this book and its materials under a Creative Commons Attribution – NonCommercial – ShareAlike 4.0 International License. I want these programs to be available to view and tweak to reach more people. Visit Creative Commons at http://creativecommons.org/licenses/by-nc-sa/4.0/ for more information. If you have any questions about this license, please reach out to me.

Hopefully, this book sparked an interest in learning more about Python. This book is merely a crash course in Python's usefulness in engineering applications. Python already has great libraries for creating GUIs, machine learning, logging, website development, pattern matching with regular expressions, interacting with data sources, and much more. I highly recommend reading through the NSA tutorial at this book's GitHub page to learn more about how Python can be useful in a variety of applications.

Contact:

- Email: pythonforengineering@gmail.com
- GitHub: https://www.github.com/alexkenan/pymae
- Website: https://www.alexkenan.com/pymae/

What did you think?

Please rate and review this book on Amazon!

https://www.amazon.com/dp/1736060619/

About the Author

Alex Kenan is a Senior Analyst at Delta Air Lines in the SkyMiles loyalty program, focusing on data analysis and program strategy. He uses Python, SAS software suite, and Oracle Hyperion to build models of future customer behavior, creates consumer insight from past customer behavior, and performs ad-hoc analyses for frequent flyer population impact. He provides instruction to new hires on data analysis in SAS software suite, Oracle Hyperion, Microsoft Access, and Python. He also has used Python to catch and prevent fraud in the SkyMiles program by identifying and blocking fraudulent redemptions and identifying hacked customer accounts.

Alex previously worked as a Propulsion Engineer for Delta TechOps where he oversaw a jet engine portfolio of nearly 200 Rolls-Royce BR-715 engines, and nearly 900 CFM56-5A, CFM56-5B, and CFM56-7B engines before that. He was responsible for the overall engine reliability for in-flight shutdown rate, unscheduled engine removal rate, and delays attributed to engine issues. He used Python to correlate engine component failure with FADEC faults which improved engine component troubleshooting and reduced engine component repair and overhaul costs.

Alex holds a Bachelor of Science degree in Aerospace Engineering from the University of Virginia and a Master of Business Administration from Emory University's Goizueta Business School.

Legal Stuff

For proper attribution to Python, the cover page photographs, and the libraries that were used.

Python Framework

The Python language and Python logos are copyrighted by the Python Software Foundation (PSF). Copyright © 2001-2020 Python Software Foundation; All Rights Reserved. See more at https://docs.python.org/3/license.html.

Cover page

Left to right, top to bottom:

1. Two Spiral Galaxies. Image Number: PR99-41.

    https://www.flickr.com/photos/nasacommons/9467312978/

2. Author's picture.

3. Ed White, First American Spacewalker. Image number: S65-30431 taken on June 3, 1965. https://www.flickr.com/photos/nasacommons/9457842193/

4. AFRC2016-0073-80 https://images.nasa.gov/details-AFRC2016-0073-80

BeautifulSoup4

BeautifulSoup is distributed under an MIT License (MIT).

https://pypi.org/project/beautifulsoup4/

Matplotlib

Matplotlib is distributed under a BSD license based on the PSF license.

https://matplotlib.org/users/license.html

MATLAB

MATLAB is a trademark of The MathWorks, Inc.

https://www.mathworks.com/company/aboutus/policies_statements/trademarks.html?s_tid=gf_trd

Numpy

Numpy is distributed under an OSI Approved (BSD) license.

https://pypi.org/project/numpy/

Openpyxl

Openpyxl is distributed under an MIT License (MIT). https://pypi.org/project/openpyxl/

Pillow

Pillow is distributed under a Historical Permission Notice and Disclaimer (HPND) license. https://pypi.org/project/Pillow/

Pint

Pint is distributed under a BSD license. https://pypi.org/project/Pint/

Pweave

Pweave is distributed under a BSD license. https://pypi.org/project/Pweave/

PyAstronomy

PyAstronomy is distributed under an MIT License (MIT Licence).

https://pypi.org/project/PyAstronomy/

Requests

Requests is distributed under an Apache Software License (Apache 2.0).

https://pypi.org/project/requests/

Tkinter

Tkinter is distributed under the PSF license.

Rolling-Shutter-Effect-Python

This program is distributed under a Creative Commons Zero 1.0 Universal license.

https://github.com/dafarry/rolling-shutter-effect-python/blob/master/LICENSE

Printed in Great Britain
by Amazon